# ANIMALS ON DISPLAY

**ANIMALIBUS** VOL. 3
OF ANIMALS AND CULTURES

Nigel Rothfels and Garry Marvin
GENERAL EDITORS

ADVISORY BOARD:
Steve Baker
*University of Central Lancashire*

Susan McHugh
*University of New England*

Jules Pretty
*University of Essex*

Alan Rauch
*University of North Carolina at Charlotte*

*Books in the Animalibus series share a fascination with the status and the role of animals in human life. Crossing the humanities and the social sciences to include work in history, anthropology, social and cultural geography, environmental studies, and literary and art criticism, these books ask what thinking about nonhuman animals can teach us about human cultures, about what it means to be human, and about how that meaning might shift across times and places.*

Other titles in the series:

Rachel Poliquin, *The Breathless Zoo: Taxidermy and the Cultures of Longing*

Joan B. Landes, Paula Young Lee, and Paul Youngquist, eds., *Gorgeous Beasts: Animal Bodies in Historical Perspective*

# ANIMALS ON DISPLAY

THE CREATURELY IN MUSEUMS, ZOOS,

AND NATURAL HISTORY

Edited by Liv Emma Thorsen,
Karen A. Rader, and Adam Dodd

The Pennsylvania State University Press
University Park, Pennsylvania

Funding for this project was provided by the Norwegian
Research Council, Grant 187858.

Library of Congress Cataloging-in-Publication Data

Animals on display : the creaturely in museums, zoos, and
natural history / edited by Liv Emma Thorsen, Karen A.
Rader, and Adam Dodd.
    p.    cm.—(Animalibus: of animals and cultures)
Summary: "A collection of essays on the historical
representation and display of animals. Using examples
from the eighteenth century to the present, the essays
situate case studies in historical and sociocultural
context while addressing the importance of visibility
for the arrangement and sustenance of human-animal
relations"—Provided by publisher.
Includes bibliographical references and index.
ISBN 978-0-271-06070-5 (cloth : alk. paper)
ISBN 978-0-271-06071-2 (pbk. : alk. paper)
1. Zoological specimens—Exhibitions—Social
aspects—History.
2. Natural history museums—Exhibitions—History.
3. Human-animal relationships—History.
I. Thorsen, Liv Emma, editor of compilation.
II. Rader, Karen A. (Karen Ann), 1967– , editor of
compilation.
III. Dodd, Adam, 1976– , editor of compilation.
IV. Series: Animalibus.

QL71.A1A55 2013
590.75—dc23
2013012122

# CONTENTS

*List of Illustrations* / vii

*Acknowledgments* / ix

Introduction: Making
Animals Visible / 1
*Adam Dodd, Karen A. Rader,
and Liv Emma Thorsen*

PART I / PRESERVING

1  Six Monstrous Pigs: Animal
Monsters and Museum Practices
in the Eighteenth-Century El Real
Gabinete de Historia Natural / 15
*Lise Camilla Ruud*

2  The Frames of Specimens:
Glass Cases in Bergen Museum
Around 1900 / 37
*Brita Brenna*

3  Preserving History: Collecting
and Displaying in Carl Akeley's
*In Brightest Africa* / 58
*Nigel Rothfels*

PART II / AUTHENTICATING

4  The Pleasure of Describing: Art and
Science in August Johann Rösel
von Rosenhof's *Monthly Insect
Entertainment* / 77
*Brian W. Ogilvie*

5  Images, Ideas, and Ideals:
Thinking with and about
Ross's Gull / 101
*Henry A. McGhie*

6  A Dog of Myth and Matter:
Barry the Saint Bernard in
Bern / 128
*Liv Emma Thorsen*

PART III / INTERACTING

7  Popular Entomology and
Anthropomorphism in the
Nineteenth Century:
L. M. Budgen's *Episodes
of Insect Life* / 153
*Adam Dodd*

8  Interacting with *The Watchful
Grasshopper*; or, Why Live Animals
Matter in Twentieth-Century
Science Museums / 176
*Karen A. Rader*

9  Polar Bear Knut and
His Blog / 192
*Guro Flinterud*

*About the Contributors* / 215

*Index* / 217

# ILLUSTRATIONS

I.1  Hall of Biodiversity, designed by Ralph Applebaum / 2

1.1  Drawing and description of a monstrous pig by Joseph Domingo Lucasa / 17

2.1  A beaked whale outside Bergen Museum arriving from Nordfjord in 1901 / 43

2.2  The Whale Hall at Bergen Museum, c. 1895 / 51

2.3  Interior from Bergen Museum, ca. 1925 / 53

3.1  Collecting Somali wild asses, May 1896 / 65

3.2  Carl Akeley with a leopard, August 1896 / 67

3.3  Kindergarten children in front of *The Fighting Bulls*, September 1954 / 70

3.4  Conservation of *The Fighting Bulls*, September 2003 / 72

4.1  Large Tortoiseshell butterfly / 86

4.2  Cockchafer / 88

4.3  Engraved title page to Rösel, *Insecten-Belustigung*, volume 2 / 91

5.1  "A room, where we prepared bird skins," photograph by S. A. Buturlin / 107

5.2  Study skin of Ross's Gull collected during the International Polar Year expedition to Point Barrow / 108

5.3  Photograph of Ross's Gulls by Fridtjof Nansen, doctored for "scientific" presentation / 117

5.4  Illustration of Ross's Gulls showing the *Fram* in the background / 118

6.1  The oldest picture of the mounted Barry in a glass case, 1883 / 134

6.2  Barry before new mounting, 1923 / 141

6.3  Barry after Ruprecht's new mounting, 1923 / 142

6.4  Reconstruction of Barry's head, 2000 / 143

7.1  The alluring gilded cover to L. M. Budgen's *Episodes of Insect Life* / 159

7.2  Title page to all three volumes of Budgen's *Episodes of Insect Life* / 161

7.3  Frontispiece to first volume of Budgen's *Episodes of Insect Life* / 162

7.4  Acheta asleep at the desk, after having devised *Episodes of Insect Life* / 168

8.1  Sketch of *The Watchful Grasshopper* exhibit in the *Exploratorium Cookbook II* / 182

9.1  Polar bear Knut / 193

9.2  Knut and zookeeper Thomas Dörflein / 205

# ACKNOWLEDGMENTS

Every book incurs for its authors and editors many debts, but those for a multiyear, multidisciplinary, internationally collaborative anthology like this one are inherently numerous and long-standing.

The Norwegian Research Council generously funded the project "Animals as Things, Animals as Signs" (2008–12), the incubator for all the seminar discussions, writing, and public outreach work at the heart of this book. The University of Oslo's Institute of Cultural Studies and Oriental Languages hosted our meetings and visiting scholars, as well as provided important administrative assistance for the project. Kristine Knudsen of the University of Oslo was especially crucial for the logistics of the *Animal Matters* exhibit.

Several University of Oslo faculty were especially helpful in giving useful advice and support, in particular Arthur Sand, Kirsten Berrum (both from the Faculty of Humanities), and Kristin Asdal (from the Centre for Technology, Innovation, and Culture). Other scholarly inspiration and assistance came from Bryndís Snæbjörnsdóttir and Mark Wilson, who worked together with the Oslo-based project team to create and mount the *Animal Matters* exhibit based on this anthology's themes and contents; Samuel Alberti, who headed up various conferences and writing projects that were both intellectually and logistically important for developing the current volume (in fact, it was at one of them that many of our principal contributors first met); and, last but certainly not least, the two anonymous scholarly reviewers of this volume, whose useful feedback improved its organization and contents.

Finally, we wish to thank the Pennsylvania State University Press, especially editor in chief Kendra Boileau, for her unwavering interest and support, and editorial assistants Stephanie Lang and Robert Turchick.

# Introduction: Making Animals Visible

*Adam Dodd, Karen A. Rader, and Liv Emma Thorsen*

Visitors entering the American Museum of Natural History's "Hall of Biodiversity" are immediately presented with a 100-foot-long wall containing over a thousand objects, most of which are animal specimens and models (fig. I.1)—but what, exactly, do these visitors see? Exhibit designer Ralph Applebaum sought to combine emotion and education in the display. "The goal," he wrote, "was to have people love the wall, love what they see, and love nature, and then confront them with what we do to nature, asking them 'why are we so cruel to it.'"[1] Indeed, shortly after its opening in 1998, a reviewer pronounced the hall a "stunning visual and intellectual design" that "instructs by seduction," demonstrating "the intricately interrelated beauty of life on earth."[2] But can this kind of looking at animals—in museum dioramas, zoo displays, and film or television programs—ever be taken at face value? Two decades ago, John Berger condescendingly declared all such modes of envisioning animals to be "compensatory," reflecting how marginalized animal lives have become in late capitalist societies.[3] Rephrasing the original question, then: What does it mean to say humans *see anything* about animals on display?

Ways of representing animals are crucial to ways of thinking about, and ultimately interacting with, animals themselves.[4] M. Norton Wise has claimed that "much of the history of science could be written in terms of making things visible—or familiar things visible in new ways."[5] Likewise, for urban, industrialized societies, practices of making animals visible have become essential to the diverse relationships human beings in those societies have formed with

FIG. 1.1 American Museum of Natural History, Hall of Biodiversity, designed by Ralph Applebaum. Photo © Peter Mauss / Esto. All rights reserved.

nonhuman animals. Although nonhuman animals exist independently of our beliefs about them, it makes little sense to speak of them in ways that attempt to locate them outside human culture. As Elizabeth Atwood Lawrence has observed, "Whenever a human being confronts a living creature, whether in actuality or by reflection, the 'real-life' animal is accompanied by an inseparable image of that animal's essence that is made up of, or influenced by, preexisting individual, cultural, or societal conditioning."[6]

Within the humanities, ways of attending to, or even speaking about, the "real animal" run the risk of becoming terminally problematized. The collected essays in this volume do not attempt to solve this dilemma, but instead map out specific ways in which real animals have become inseparable from a variety of human modes and practices of display: museums, illustrated books, even the Internet. Although it may seem, Berger famously observed, that "in the last two centuries, animals have gradually disappeared,"[7] we suggest that human-animal relationships have continued to become increasingly complex, in ways that are not always attributable to the so-called disappearance of animals. Certainly, we can consider the last two centuries as a period in which animals have been removed from the daily lives of many,[8] but just as certainly can we see this period as one in which many have been removed from the daily lives of animals themselves. Regardless of the direction it has taken, this undeniable alteration of a historical proximity to animals has been the impetus for a plethora of representational practices that, broadly conceived, work to fill in the gap between humans and animals—to help "bring us closer," not necessarily to the animal itself but to the animal as imagined within a historio-cultural space.[9] Ironically, many of these productions may ultimately work to exacerbate the very nostalgia, distance, and ignorance they were devised to remedy. What is certain, however, is that these productions represent an ongoing process of making animals visible.[10]

Emerging from an international multidisciplinary research project based at the University of Oslo, "Animals as Objects and Animals as Signs," this collection represents a range of responses to a set of key questions about animal materiality and display that cross museum studies, cultural history, history of science, and environmental studies: How are animals transformed into objects? How do animals become signs? How do aesthetics intersect with the standardization of animals? What are the ideological, cultural, and pragmatic dimensions of these processes?

If we encounter animals as objects or signs more often than we encounter animals themselves—or even if we have mournfully lost the ability to conceive of an authentic and direct animal encounter—this surely says something about our inability to live in any way without animals. We need their bodies as much as we need our own thoughts about their bodies, images of their bodies, models of their bodies. Animal bodies are intrinsic to the human imagination; indeed, the oldest surviving images produced by human beings, rendered some twenty-six thousand years ago, are images of animals. Yet it is the ongoing uncertainty, particularly pressing in these times of enhanced concern for the natural environment, of how we should think about and interact with animals that informs the essays collected here, all of which are motivated by a desire to augment our understanding of the animals, real and imagined, that surround us from infancy until old age. Just as the infant's attitude toward animals will differ from the octogenarian's, so, too, do cultural attitudes toward animals change over time as knowledge about them is gained, lost, recovered, reshaped, and reconstructed. But if, as Lorraine Daston and Gregg Mitman suggest, the "how" and "why" of thinking with animals deserves greater attention, so, too, do the relations between animal materiality and animal representation.[11] For even while this materiality is often modified, distorted, obscured, or erased across a range of representational practices and institutions, it nevertheless underpins human relationships to animals themselves.

Such observations, far from exclusively inhabiting the lofty heights of philosophical discourse, are of direct relevance to both animals themselves and the publics that are literally constructed around their display. This was a key concern of an art exhibition devised as part of the "Animals and Objects and Animals as Signs" project at the University of Oslo. Designed by Bryndís Snæbjörnsdóttir and Mark Wilson, and titled *Animal Matters*, the exhibit addressed some of the diverse and often problematic methods of representing animals that have developed since the eighteenth century. Including drawings, photographs, sculptures, motion pictures, preserved animal bodies, and scientific instruments, it sought not merely to represent animals, but to present some of the ways in which animals are themselves represented—to draw attention to representational practices as they manifest across a range of sites and contexts, including the art gallery itself.

The institutional sites and spatial contexts examined by authors in the present volume served as a preliminary point of departure for the design of the *Animal Matters* exhibit. In some cases, such as zoos and natural history museums, the materiality of the animal was foregrounded and privileged to emphasize how it could come to define the animal itself, even when enclosure fences, Plexiglas covers, or glass cases frame human interactions with it. In other instances, such as illustrations and photography, the animal's presence was interpreted as at once more one dimensional and more spectral. Within the modern tradition of Western natural history, a strong emphasis on realism has oriented representations of animals according to accepted conventions of what is most lifelike—in short, animals as they really are. Yet upon closer scrutiny, it becomes evident that what is ultimately presented are representations of animals as they really are for the producers of the representations themselves, for even while these representations aspire to empiricism and objectivity from the perspective of a detached observer, they fail to escape their socio-historical context. Paradoxically, African elephants have outlived all efforts to hunt, preserve, and display them in vaunted galleries—so they have lives as well as after-lives[12]—but what remains on the Field Museum floor in Chicago, for example, is as much a biography of elephants as it is of their creators.

As scholars, we are acutely aware that this paradox applies to us, too—and so we have organized this volume in ways that draw attention to it and demand further reflection. We have grouped the essays according to what we see as a set of recurring practices used to make animals on display visible—preserving, authenticating, and interacting—in order to showcase how these practices are both firmly anchored to specific historical projects and moments *and* dynamically relocating across and between them.

Questions of cultural values and biological identity are acute in the projects of understanding *specific animal bodies* in *specific contexts*: a new specimen of bird in a museum drawer, a captive animal in a zoo or presented on a blog, an infamous animal display in a natural history collection. At the same time, these animals become iconically visible only through occupying multiple and contradictory cultural and temporal spaces. Taking the animals described here as both unique instances and embedded in a larger, interconnected history of representational practices, we suggest, illuminates a variety of critical aspects inherent to our own conceptions and understanding of what animals are, what they mean, and what they should and should not be "used for."

## Preserving

Lise Camilla Ruud opens the volume with an examination of how six preserved monstrous pigs became objects of value and exchange in the museum culture of eighteenth-century Spain. As historians of natural history have recognized, "monsters" (deformed animals) served an essential function in the grand project of normalization, rationalization, and disenchantment that typified the Enlightenment. An emergent, modern scientific culture, oriented by a new and emboldened empiricism, sought to reveal the fundamental order of Nature as a way to make sense of its often overwhelming diversity. Integral to this process was the ability to demonstrate exceptions to the "rules" that governed the morphology of normally developed animals. Ruud shows that the emphasis of monstrous difference ultimately served an overarching agenda that stressed the conformity of nature to a visible plan. But the monstrous pigs did more than this; they also became commodities, gifts, and gestures, elevating the social status of their donors and recipients, many of whom were eager to associate themselves with the prestigious world of institutionalized natural history. Each of Ruud's pigs has its own story, and each is an explicitly unique individual. Yet, taken together, they can be seen as the unlikely cast of a particular chapter of eighteenth-century Spanish history: dead and preserved in glass jars, yet instrumental in a nexus of human relationships, and indeed reverberating with other treatments of animal monstrosity throughout Enlightenment Europe.

Moving into the nineteenth century, Brita Brenna turns attention to the emergence of the glass case (or vitrine) and its impact on the preservation, display, and reception of museum objects—including taxidermied animals. Her primary focus is on how the localized collection and display practices of a relatively small and unknown natural history museum signified wider ambitions of universalized methods—a turn-of-the-century instance of "thinking globally, acting locally," as it were. Bergen Museum, on the western coast of Norway, was eager to participate in the burgeoning museum scene of the late nineteenth century, perhaps most illustriously exemplified in Europe by the British Museum (Natural History). Central to the nineteenth-century museum was glass—today a ubiquitous material routinely taken for granted, but during the nineteenth century both a source of novel fascination and a substance enabling a variety of previously impossible practices. Beyond the specific glass case of a particular display (as discussed by Thorsen in this volume), Brenna emphasizes the general importance of glass for allowing displays of rare and valuable objects in full view of the public without the direct assistance or instruction of museum staff. Often, however, any specific stuffed animal behind or within glass, while visually arresting, was envisioned by museum curators as merely an illustration of the accompanying authoritative text explaining its taxonomy, habits, habitat, and so on. Although Brenna foregrounds the historical example of Bergen Museum, her essay operates against a more nebulous backdrop, in which the subtle intangibility of glass, and its profound effects on how we look (through it) at objects and things, shapes a modern visual culture and its complex, problematic inclusion of the animal subject, linking up with, for example, the monstrous pigs of El Real Gabinete de Historia Natural discussed by Ruud.

Nigel Rothfels takes up the various material and interpretive problems inherent in preserving animals through an exploration of the work of Carl

Akeley. Akeley—widely regarded as one of the most accomplished taxidermists of all time, with numerous works housed and displayed at the Field Museum in Chicago and the American Museum of Natural History in New York—was working at the turn of the nineteenth century, a time when the human impact on the prosperity and indeed survival of many nonhuman animals was beginning to become apparent. Rothfels engages directly with the central paradox of Akeley: motivated by the desire to prevent many of the most majestic species of mammals from going extinct, Akeley undertook to hunt, kill, and stuff them for the benefit of science and the general information of the public. Unlike numerous contemporary hunters, for whom displays of bravado, conquest, and the thrill of the chase were almost all consuming, Akeley was mournful of the steady disappearance of animals, suggesting a particular context for his taxidermy. Akeley's stuffed animals are not so much evidence of domination and defeat (i.e., trophies), but rather can be read as poignant signs of the fragility of the species embodied by the particular specimens displayed. Rothfels shows how, in particular, an informed reading of Akeley's memoir can shed further light on his taxidermy—reminding us that the "meaning" of the animal (or its effigy) is always shaped by a diversity of contingent factors, and that in some cases, relationships that form between humans and animals may constitute legitimate paradoxes that cannot be easily (or perhaps ever) resolved.

## Authenticating

Brian Ogilvie examines how art and science intersected around very small animals—insects—in the eighteenth century. Early modern and Enlightenment studies of insects faced the challenge of making animals visible in a unique way, since their subjects are often too small to be seen, or seen well, with the unassisted eye. This meant that standardized forms of magnification—through the lens and on the page—were necessary for the production of authentic portrayals. These methods did not develop overnight, but rather emerged over time; indeed, they are still developing, and continue to marry aesthetic concerns with pragmatic requirements. As most historians now recognize (though it bears repeating nevertheless), art and science were not considered mutually exclusive disciplines during the Enlightenment period; indeed, the pursuit of one required the skills of the other, especially when it came to studying and representing insects. The need to establish some validity for the study of insects inspired strikingly detailed illustration, a practice that began to gain considerable momentum during the eighteenth century with images as beautiful as they were accurate. August Johann Rösel von Rosenhof, the central figure of Ogilvie's essay, emerges during this time as a significant proponent of not just what would later be labeled as the science of entomology, but what might be called an insect aesthetic. We find here an exemplary case of art, science, and theology converging not only in the animal, but also in its "world." The aesthetics of butterflies and beetles that Rösel developed simultaneously authenticated the "lifeworlds" of the insects he observed and the methods used to observe them. Insects were to be understood (and portrayed) both as specimens detached from their unfamiliar, subvisible "world," and as emblematic of that world itself, which was becoming increasingly accessible to those with the technology and natural history skills required to explore and document it.

Few material practices have changed both natural history and animal-human "lifeworlds" more than exploration, so accordingly, Henry McGhie turns to a case study of one particular animal—Ross's Gull—to demonstrate these important historical shifts. McGhie considers the problem of what counted as knowledge of this rare bird—was it enough to see a well-preserved specimen in a museum drawer, or was it necessary to travel to the Arctic and encounter the animal in its natural environment? McGhie shows how the debate around the scientific naming of animals reflected new modes of engagement: that this bird was later referred to by association with its discoverer, the quintessential Polar explorer James Ross, rather than by its physical characteristics (such as its trademark rosy breast), pointed to different fields of activity in which particular animals retained (or, in some cases, lost) their global "reputation." Ross's Gull, McGhie concludes, provides a powerful example of how scientific and cultural practices collaborate to transform animals from complex, multiple creatures into standardized, singular idealizations.

Liv Emma Thorsen confronts the fact and artifact of Barry, the most famous Saint Bernard rescue dog. Thorsen finds the stuffed Barry, on display in the Natural History Museum of Bern since 1814, to be materially emblematic of the "Barrylore" surrounding the animal. Thorsen shows how, through the practices of taxidermy, the natural and cultural history of the Saint Bernard intersected—and in turn, how this museum specimen came to represent both the iconic "faithful dog" and the representative type of its breed. Literally disembodied, with the skull and skin of "the real" Barry now housed in different parts of the museum, Barry's display nevertheless retains an aura of authenticity. Because (rather than in spite) of how this animal (now behind glass) materializes the imagined Saint Bernard, it continues to inspire modern natural history museum visitors.

## Interacting

But what counts as natural history, Adam Dodd reminds us, is dynamic and contextual—so there is more to the popular nineteenth-century accounts of insects generated by L. M. Budgen than might first appear to a modern reader. In the representational conventions of these stories, Dodd finds the origins of common anthropomorphic portrayals of insects—but also, a different mode of knowledge making, one that Budgen saw herself as creating with, rather than transplanting onto, animals. Budgen's insect books, much like Rösel's more than a hundred years earlier, navigated complex boundaries between subjective engagement with and objective distancing from nature—and from them, as with all popular natural history, their contemporary audiences drew moral and allegorical truths. To the extent that we can see these insect stories as constituting, rather than constructing, a "lifeworld" (and in that sense, as representing a continuation of the conventions visible in Rösel, as examined by Ogilvie in this volume), Dodd suggests we can resist static "mechanomorphic" interpretations of animal-human relations. Perhaps above all, Budgen's unique authorial mode encourages, and allows, imaginative and fruitful interaction with living insects, beyond the pages of the book.

Karen Rader's essay on *The Watchful Grasshopper* brings us to the twentieth century in order to consider the live animal's uneasy place within the educational agenda of the interactive science museum. The grasshopper, a human agricultural pest, here elicits empathy as the living subject of a scientific experiment controlled by museum visitors. Rader's examination of this peculiar and largely overlooked case of insect-human interaction illustrates another chapter in a long and complicated history of cultural mediations of insects that evoke (often unsettling) emotions in the human observer; there are connections, for example, with the insects of bygone centuries discussed in both Ogilvie's and Dodd's essays. Intended as a public educational installation dealing with the insect's visual perception, the exhibit aroused a largely negative response from museum visitors who, as if suddenly provoked to ask themselves "What is it like to be an insect?," protested against the inhumane treatment of an anonymous arthropod. *The Watchful Grasshopper* draws attention to the important role that representational conventions play in mediating human-animal relationships; the ways in which we think, and indeed feel, about animals are very often responses to how our expectations intersect with the ways animals themselves are presented to us for consideration. For this reason, Rader explains, live animal displays in museums have both "transgressive and contradictory possibilities."

Finally, Guro Flinterud documents how an older form of animal-human relations, that between a zookeeper and his or her animal charges, has been transformed by new media—specifically, the Internet. The polar bear Knut, born in 2006 at the Berlin Zoo and abandoned by his mother, was initially presented to the world as a seemingly endless stream of images: the cute cub, being hand-reared by a human (male) caregiver. Within this context, Flinterud examines how Knut was transformed in cyberspace through the narratives of blogging. Interactions between Knut (now represented as a blogger himself) and his online fan community (who constructed themselves around this mythical animal representation) made explicit the contradictory cultural values and opinions connected to the polar bear in the early twenty-first century. Such attention, ultimately, was not enough to save Knut—he died at the age of four, collapsing into a pool of water in his enclosure, surrounded by zoo visitors, and within forty-eight hours, amateur video of his death appeared online. That this particular animal mattered—to many humans—seemed never to be in question. But Flinterud's analysis, like the others in this volume, encourages us to tend better to how animal representations both shape and connect us to the living, breathing, thinking animals behind such representations—that is, to the very matter of the animal itself.

## NOTES

1. As quoted in Pes, "A World Vision," 32.
2. "In the Hall of Biodiversity."
3. Berger, "Why Look at Animals?" 26.
4. Haraway, *Primate Visions*; McHugh, *Animal Stories*.
5. Wise, "Making Visible," 75.
6. Lawrence, "The Sacred Bee, the Filthy Pig, and the Bat out of Hell," 302.
7. Berger, "Why Look at Animals?" 11.
8. Lippit, "From Wild Technology to Electric Animal."
9. Geographers have been especially attentive to these spaces. See, for example, Wolch and Emel, *Animal Geographies*; and Davies, "Caring for the Multiple and the Multitudes."
10. For a thoughtful reflection on the relationship between visual and narrative representations in response to Berger's claim, see O'Brien, "The Pig Stays in the Picture."
11. Daston and Mitman, "Introduction."
12. Alberti, *Afterlives of Animals*.

BIBLIOGRAPHY

Alberti, Samuel. *The Afterlives of Animals*. Charlottesville: University of Virginia Press, 2011.

Berger, John. "Why Look at Animals?" In *About Looking*, 1–28. New York: Pantheon, 1980.

Daston, Lorraine, and Gregg Mitman. "Introduction." In *Thinking with Animals: New Perspectives on Anthropomorphism*, edited by Lorraine Daston and Gregg Mitman, 1–14. New York: Columbia University Press, 2005.

Davies, Gail. "Caring for the Multiple and the Multitudes: Assembling Animal Welfare and Enabling Ethical Critique." Paper presented at the "Sentient Creatures" workshop, Oslo, September 2010.

Haraway, Donna. *Primate Visions*. New York: Routledge, 1990.

"In the Hall of Biodiversity." *New York Times*, June 1, 1998.

Lawrence, Elizabeth Atwood. "The Sacred Bee, the Filthy Pig, and the Bat out of Hell: Animal Symbolism as Cognitive Biophilia." In *The Biophilia Hypothesis*, edited by Stephen R. Kellert and Edward O. Wilson, 301–41. Washington, D.C.: Island Press, 1993.

Lippit, Akira Mizuta. "From Wild Technology to Electric Animal." In *Representing Animals*, edited by Nigel Rothfels, 119–36. Bloomington: Indiana University Press, 2002.

McHugh, Susan. *Animal Stories: Narrating Across Species Lines*. Minneapolis: University of Minnesota Press, 2011.

O'Brien, Sarah. "The Pig Stays in the Picture: Visual/Literary Narratives of Human-Animal Intimacies." *Reviews in Cultural Theory* 3, no. 1 (2012), http://www.reviewsinculture.com/?r=77 (accessed September 8, 2012).

Pes, Javier. "A World Vision." *Museum Practice Magazine*, Spring 2004.

Wise, M. Norton. "Making Visible." *Isis* 97, no. 1 (2006): 75–82.

Wolch, Jennifer, and Jody Emel, eds. *Animal Geographies: Place, Politics, and Identity in the Nature-Culture Borderlands*. New York: Verso, 1998.

# PART I

PRESERVING

# Six Monstrous Pigs: Animal Monsters and Museum Practices in the Eighteenth-Century El Real Gabinete de Historia Natural

*Lise Camilla Ruud*

In 1774, a monstrous piglet, preserved in alcohol within a glass bottle, was delivered to El Real Gabinete de Historia Natural in Madrid, the first public museum in Spain. The deformed animal was a gift from Gaspar Soler, governor of the village of Almadén, and was one of the very first monsters to be included in the cabinet's collections. At the beginning of the 1770s, the Gabinete's collection of animal monsters consisted of just a few disproportioned, malformed specimens—but by the turn of the century, around two dozen had been donated to the museum.[1] A large proportion of the monsters were domesticated animals, such as calves, chickens, cats, dogs, and pigs, but occasionally some wild animals, such as fish or reptiles, were also included. Among the collected animal monsters were six monstrous pigs, all offered to the museum between 1774 and 1798. Most of them were malformed fetuses, such as the piglet with one head and eight legs, or the one with an elephant-like trunk and only one eye. Archival documents about the animal monsters reveal that they were integrated into a broad variety of museum practices and that they were represented in multiple ways.

Normalization, rationalization, domestication—key themes in the research on eighteenth-century monsters—are closely connected to the idea of the Enlightenment as an age of *Entzauberung*, or disenchantment: of reducing the unknown by explaining the natural world through empirically based knowledge, and of establishing order and rationalizing the world through the classification of the natural world itself.[2] While monsters in the Renaissance, among the learned, were considered portents and results of a playful and astute nature, eighteenth-century naturalists came to consider them parts of a regular order of nature, as rationally explicable phenomena. This process is often seen in correspondence with the establishing of teratology, the biological study of birth defects, as a subdiscipline within medicine in the early nineteenth century.[3] Over the last decade, a growing number of articles and books have questioned and challenged the field's emphasis on rationalization, arguing instead in favor of a more complex, historically specific narrative of eighteenth-century monsters.[4]

The eighteenth century witnessed a significant increase in the representation of monsters, argues Michael Hagner; therefore one should take a closer look at the spaces of representation and the integrative practices where "monsters were pushed to get in."[5] Andrew Curran and Patrick Graille describe the eighteenth-century monster as "a fluctuating beast" and point to the lack of a coherent definition of the monstrous within Enlightenment thought.[6] Lorraine Daston and Katharine Park argue that monsters and marvels should be studied as nonlinear and nonprogressive cultural phenomena.[7] Yet Daston and Park, in demonstrating how monsters and marvels were "exiled to the hinterlands of vulgarity and learned indifference" during the eighteenth century,[8] continue to tell a story that follows the rise and decline of the monstrous in relationship to science, while leaving out, according to Knoppers and Landes, a broad range of metaphorical and political uses of the monstrous.[9]

This essay follows, in part, Daston and Park's argument about the decline of interest in the monstrous within the context of eighteenth-century science. But it also explores institutional, political, and personally motivated meanings of monsters. The close study of the multiple ways of representing six monstrous pigs at El Real Gabinete de Historia Natural in Madrid, then, adds further nuances to a historical understanding of the multiple meanings of animal monsters in the eighteenth century, and suggests how such historical specificities came about. The pig monsters—through being offered as gifts to the museum, being described in texts or represented in images, and being displayed—served to establish and situate the cabinet within the Enlightenment context of museum and science. While these Madrid monsters indeed seemed to have been domesticated and rationalized, a closer look at the processes through which they became so reveals their importance for various museum practices at the public Royal Cabinet. None of the six monstrous pigs discussed in this essay was represented in images; only textual documents

FIG. 1.1 Drawing and description of a monstrous pig made by Joseph Domingo Lucasa in 1781. España, Ministerio de Cultura, Archivo General de Indias, MP-Estampas, 206.

remain to tell about their existence. The image (fig. 1.1) made by Joseph Domingo Lucasa, of a monstrous pig seen in the Philippines, was directed to the minister José de Gálvez in Spain, and destined to the Royal Cabinet in Madrid by the Spanish governor in Manila, Jose Basco y Vargas. The etching, however, never reached the cabinet; probably it was sent to the General Archive of the Indies in Seville instead, where it is still kept.[10]

## A Tiny Monstrous Pig "Worthy of Being Displayed at the Royal Cabinet"

The Spanish Royal Cabinet of natural history was established in 1771, when Carlos III acquired the collections of the Ecuadorian Creole Pedro Franco Dávila, tradesman in cocoa and reputed private collector, until then residing in Paris. In return for donating his collections to the Spanish crown, Dávila became the first director of the museum—a position he held until his death in 1786. Several central Spanish Enlightenment thinkers had urged the crown to establish a cabinet of natural history. Not only would this be useful for enlightening the population and giving glory to the king and *la patria*, but it would also help Spain keep pace with European progress, it was argued. The royal decision to establish a public cabinet in 1771 was one in a series of Bourbon reforms.[11]

In 1774, when Gaspar Soler donated the first piglet monster, the museum's collections were stored in the basement of the royal palace in the Retiro, waiting for the Goyeneche Palace in calle Alcalá (purchased in 1773) to be reconstructed into a proper Enlightenment museum. The palace was remodeled in order to house and display the collections in nine halls, and the cabinet opened its doors to the public in November 1776. The museum was open for everyone to visit, every Tuesday and Thursday from morning to afternoon—the only condition governing who could attend was that the visitors needed to be decently dressed.[12] Outside the opening hours, however, the director would receive a more exclusive audience, such as members of the royal family and court, high officials, aficionados, nobility, and foreign travelers—some of whom came to be important contributors to the museum.

At the beginning of the decade, institutional rules had not yet been entirely defined or set, and questions concerning how the museum should educate its visitors, or which persons should participate in the production of knowledge, were subject to negotiation and debate. Letters about the new museum were sent all over the empire to inform high officials such as viceroys, governors, and military officers as well as clergy and nobility. Officials in the Spanish American and Asian colonies were given instructions to send the museum a huge variety of objects.[13] Dávila engaged actively in networks of written correspondence with aficionados, collectors, owners and directors of private and public cabinets, as well as with members of scientific academies all over Europe. In his letters he shared information about the collections, new acquisitions, and the forthcoming opening of the cabinet, to situate the Madrid museum on the European map of collectionism. As director of the Royal Cabinet, Dávila was an accepted member of various academies and societies, and he corresponded with, among others, Linnaeus.[14] The museum director envisioned that natural history, a discipline neglected for so long in Spain, would finally get the attention it deserved. The new institution, he believed, would contribute to the promotion of the natural sciences in the Spanish Empire.[15]

The cabinet would also play a central role in creating relations between king and population. A primary goal of the eighteenth-century reforms introduced by the Bourbons was to transform the population into a suited and well-disciplined public.[16] The establishment of public institutions played a major role in these efforts. The staff at the cabinet held an important intermediary position in the symbolic space between king and population. The museum management took this task seriously, and the director eagerly engaged in outreach to establish relations and connections as mentioned above. Such relationships ensured the inflow of a stream of objects and augmentation of the collection—but also guaranteed the participation of the right kind of contributors to situate the institution within the context of Enlightenment knowledge.

The piglet donated by Gaspar Soler was accompanied by a humble letter underscoring his confidence in the director's authority. Soler wrote that he himself was not sure whether the small monstrous piglet was "worthy of being displayed at the Royal Cabinet of Natural History." He wrote that he nevertheless trusted that the museum director would decide this, since Dávila had "superior comprehension" of the matter. Soler also added that he was more than happy to wait for further orders from the director concerning his future contributions to the museum, and noted that he "truly wanted to satisfy the director."[17] Not everyone was allowed to enter the royal museum milieu as a contributor of knowledge—and among the possible ways to enter this sphere, getting the director's approval would be one of the most certain avenues to pursue.

Soler's donation proposed an important relationship that connected the center of natural history knowledge-making to local governments and their elites. Soler was not only governor in Almadén, a village in Castilla la Mancha, but also the superintendent of the royal mines there.[18] He kept an influential position within the local state apparatus and presumably in the local, emerging, and enlightened bourgeois elite as well. He probably had an interest in positioning himself within the institutionalized sphere of science and knowledge established throughout the eighteenth century by the Bourbons.

Not all donations offered to the museum were given a response. This may imply that either the donation or the donator was not considered very important, or simply that the response has been lost. No response from the museum director to Soler has been kept, so it is difficult to determine the impact of Soler's donation. However, documents concerning the inclusion of another monstrous piglet clearly suggest the importance of such gifts to the museum. In August 1786, the Duke of Híjar donated a piglet monster to the museum. The duke believed it "worthy of being displayed at the Royal Cabinet, being so extraneous and particular."[19] Híjar, like Soler, moved in multiple spheres of authority and power: he was a duke and the director of a hospital in Madrid.[20] Like Soler, the Duke of Híjar can be associated with the institutionalized sphere of science and reform in contemporary Madrid. The gift was highly esteemed and the piglet was included in the collections. The responding letter sent from the museum to the duke demonstrated gratitude, and assurances were made that the gift would be mentioned to the Count of Floridablanca. Floridablanca was the chief reform minister under Carlos III, the king's right-hand man, as well as the high protector of the cabinet. The assurances implied not only that the news about the donation would reach all the way up to Carlos III; they were also a demonstration of the close link between the cabinet and the crown. The donator and the museum were connected through the gift of the monstrous pig, and thereby situated in a network of institutionalized reform and royal patronage.[21]

The piglets donated by Gaspar Soler and the Duke of Híjar seem to have attained their importance due to more than their bodily deviances. They were not primarily interesting monsters to be investigated in a scientific manner or displayed to visitors at the museum. In fact, the deformity of the piglets is not mentioned in the letters. The absence of written descriptions or memorandums about the piglets' monstrosities as scientific or didactic objects is interesting, but equally interesting is the presence of the documents that can guide us into the socio-institutional aspects surrounding the monsters. These two pig monsters were valued gifts to the king and contributions to the Enlightenment knowledge distributed in his empire, and they were particularly important due to their capacity to represent both the donor's important contributions to and participation in the museum, as well as the museum's position within the royal sphere.

### "The Rarest Character Among the Monsters"

In 1778, the museum director Dávila wrote a two-page description of a piglet that had been donated to the museum. In contrast to the letters accompanying the piglets of Soler and the Duke of Híjar, this document lacks information about the monster's provenance. The fact that only the date when it was donated is mentioned indicates that Dávila's text belonged in another context and implies that such information was of secondary importance. Here, much more than in the cases discussed above, it was the piglet and its deviances that were the focus—as an object for scientific study. In the case of human monstrosities, the practice was the opposite, in that it was of crucial importance that the phenomenon was witnessed and testified to by reliable persons.[22] In the case of animals, though, this seems not to have been equally as important.

As in many similar contemporary descriptions, a comparative approach is used in order to establish categories—and the text starts with an underscoring of the monster's normality: "Its structure is similar to that of any other pig, except for the head, which is irregular." However, the focus on normality is soon abandoned in favor of a narrow focus on the abnormalities, and the monster piglet is described as "the rarest character among the monsters" in the collections.[23] This status was gained from its three particularities: it had a nose shaped like an elephant's trunk; its two eyes were melded into one, like a cyclops; and it had a very extraneous position of the jaws and teeth, with three canines, one in the front mouth sticking out like a beak. This accumulation of particularities made the piglet the most interesting and rare among the monsters.[24]

Compared to the piglets donated by Soler and the Duke of Híjar, which I have interpreted within a socio-institutional context, the description made by Dávila draws attention to the most singular features of the piglet, and represents them within a scientific discourse familiar to the eighteenth-century naturalist. While monsters were indeed rare objects in Renaissance collections, during the Enlightenment they became more common and were no longer infrequent as collectibles. Therefore, the number of descriptions published in scholarly journals about monsters increased tremendously in the first half of the eighteenth century. As a response to the massive amount of such descriptions, scientists argued that it was not of any scientific interest to describe the same aspects over and over again, as this would make the history of monsters appear as an endless repetition of what was already known, and nothing important, substantial, and new could be learned. Instead of repeating the known, the focus should instead only be on the most singular characteristics of the monsters.[25]

To describe the monstrous by focusing narrowly on the most singular implies a particular relation between abnormal and normal. Within eighteenth-century learned naturalist societies, monsters functioned as instances of the normal rather than as examples of a wider category of monstrous, and often were regarded as experiments made by nature itself.[26] By investigating the monstrous, normality could be explored, and monsters were seen as having an analogical relationship with the normal. A focus on their singular characteristics highlighted this relationship—and the descriptions and investigations centered on the integration of the monstrous into the regularities of nature.[27] One can imagine how, for instance, the cyclopism of the rare monstrous piglet must have been eagerly discussed by aficionados visiting the cabinet: Did the eye have one or two optical nerves, if any? How could it be the result of God's will? How did it fit into an ordered nature? And the proposed answers would perhaps be focused on the fact that the two eyes were not totally melded together—it had two eyeballs, and a small portion of skin-like texture in the middle. Due to this it might have represented a privileged opportunity to investigate the borders between normal and abnormal: it was an in-between case. It was neither a pair of normal eyes nor a completely monstrous cyclops eye: so then, perhaps it could reveal something about the process of generation? It had been argued that cyclopism could not possibly be God's will because it would have been implausible to create a perfect eye without an optical nerve; if such an eye existed, then, generation became something uncertain and unforeseeable.[28] By investigating and discussing the monstrous piglet's singular eye, as an in-between case, perhaps one could get closer to an understanding of the connections between the monstrous structure and the normal.

The point here is not so much to identify the content of the scientific reflections as it is to point out that the piglet was represented with reference to a specific scientific textual genre, and hence gained importance within a strictly limited public sphere. The rare piglet, accompanied by its description, was intended for a specific type of audience, one that was familiar with scientific discourse, and one that would not be frightened or shocked by seeing the piglet. Perhaps the specimen was set aside in the closed hall where human anatomical specimens were placed, open for an invited audience only, such as members of the royal family and the court, high officials, diplomats, nobility, foreign high-society travelers, and aficionados.[29] The scientific reflections were not for everyone to participate in; they were not a common public affair. The handwritten description was not published, it did not contain any revolutionary findings, and the content would perhaps have been more interesting within an international scientific context in the first part of the century. Toward the end of the century, this method probably had been stabilized as a standard scientific approach. It had been converted into a common way of describing monsters and incorporated into a state of normal science in a Kuhnian sense.[30] By using this genre, Dávila was reproducing and maintaining this way of incorporating monsters into the socio-scientific world of the museum. The monstrous piglet, "the rarest character among the monsters," was perhaps deemed ill suited for the general visitor. It was too much of a monster, and too complex to relate to the normal, the useful, and the understandable in the contemporary culture within which the museum had its didactic functions for the broader public.

Monsters were also displayed for the lay audience, but these monsters would have a more accessible and naive monstrosity. They would be carefully selected to convey messages suitable for this group of visitors. This becomes particularly clear in the publication of the *Coleccion de laminas que representan los animales y monstruos del Real Gabinete de Historia Natural de Madrid* (1784–86), made by the museum taxidermist and anatomical painter Juan Bautista Brú y Ramon. The collection consisted of seventy-one etchings of animals and monsters in the cabinet, all accompanied by an individual description. Five of the etchings are of monsters: a chicken with three legs; a hare with two bodies; a calf with two heads; a lizard with two tails; and a calf with a monstrous head and only one eye. Seen as a whole, the set of five monsters represented in text and image combines various intentions to explain and normalize the monstrous within a didactic sphere for a lay audience: by referring to the utility of the species (as in the case of the monstrous calves and the two-tailed lizard), or by describing how even monstrous bodies functioned almost normally (as in the case of the three-legged chicken).

Certain kinds of monsters, however, were judged unsuitable for the general public to see, particularly human monstrous fetuses.[31] Even the piglet was probably too complex for those exhibitions open to the public. Such an exhibition could be more confusing than instructive and more frightening as well, as the piglet fetus preserved in alcohol may have a certain resemblance to the human fetus, being approximately the same size and with a similar skin color. Moreover, it could easily have been associated with what was seen as the vulgar superstition of the unlearned. Cyclopses have a long history as mythical, folkloric monsters—in Dávila's description, the pig was characterized in Spanish as *ojanco*, a reference to the Cantabrian one-eyed folklore figure Ojáncano, a northern Spanish variant of the cyclops embodying rage, cruelty, and evil.[32]

Why Dávila chose to use the term *ojanco* is difficult to say; however, it is reasonable to interpret his description as an expression of a scientific language not yet scientifically pure (as would be the case within nineteenth-century teratology) and at the same time cleansed neither of the earlier, Renaissance view of monsters as marvels, nor of the contemporary vulgar or common expressions.

### "Pig of Extraordinary Magnitude"

In 1779, the same year Dávila wrote about the rare piglet discussed above, Francisco de Guya, an archpriest from Úbeda in Andalusia, visited the cabinet. The archpriest told the museum director about a stuffed "pig of extraordinary magnitude" that he had in his hometown, a pig that he thought was "worthy of being displayed at this Royal Cabinet." The descriptions of the gigantic pig must have aroused the director's interest, since he openly welcomed it as a gift to the cabinet. A monstrously big pig was obviously worthy of being displayed in the Royal Cabinet. In connection with the donation, Dávila wrote a letter to Luis de Almagro, another clergyman in the same town, to sort out the practical arrangements concerning the shipping of the huge pig from Úbeda to Madrid. In the letter, Dávila demonstrated his gratitude: "In the catalog that we are making, it will be explained who donated it, the village where it came from, and if there are by chance other circumstances worth mentioning, such as its weight."[33]

What meaning did this pig and its size achieve? There are multiple possible interpretations. It is relevant to see the acceptance of the monstrous pig in relation to the importance of establishing networks of donors, as discussed in the first section. This pig was donated by an archpriest while another clergyman arranged the shipping. Due to its authority and presence all over the Spanish Empire, the clergy was an important contributor to the museum. The gift would be registered in the catalog, which was a way of expressing gratitude within the public sphere; being mentioned in the catalog was a way of being honored, even though it was not as prominent as being mentioned to the chief minister, the Count of Floridablanca, or Carlos III himself.

The magnitude of the pig appears to have been its most important feature. Its size made it a singular and rare object, and even though the cabinet displays ideally should show a classified, regular order of nature, the singular and outstanding objects were perhaps more suited to represent and convey messages about magnificence and grandeur. The cabinet was a place containing and showing a broad range of the *maravillas*, the wonders that existed in imperial nature. The cabinet was a royal cabinet—its task was not only to convey knowledge about natural history, but also to ensure that the collections incorporated and represented royal, imperial glory—and the exhibitions should therefore arouse the spectator's patriotic spirit.

On the other hand, the pig was implicitly incorporated into a normalizing practice: as a monstrously big pig, it made manifest understandings of the normal pig. Normal specimens of domestic animals would not have been given a place in the collections, but the deviant monstrous ones that were put on display were often juxtaposed with the history of the regular domestic species. However, such relations between the monstrous and the normal would soon be considered irrelevant or uninteresting. In a museum guide that was published in 1818, for instance, a huge pig is mentioned and very likely it is the same pig: "On top of the two last shelves the only thing one sees is a large pig."[34] Its size had been reduced, if not literally, then indeed symbolically, and it stood out as something odd, something without relevance: it was big, and that did not in itself qualify for being displayed some forty years after it had been valued as a marvel.

## "They Have Forgotten to Give Him the Degree of Lieutenant Colonel"

Of all the monsters offered to the cabinet, only one was explicitly rejected. The decision concerned an aborted piglet offered to the museum in 1783 by Paula Montenegro y Palomo. She was a resident in the small village of San Felices de los Gallegos in Salamanca and the wife of a military commander. The letter is particularly interesting because it is one of the few addressed to the cabinet in the eighteenth century to have been written by a woman:

> Honorable Sir, I am aware of how much you appreciate strange abortions of nature, in order to put them in your cabinet. I do not know whether that will be the case with the abortion of a sow that took place on the eighteenth of this month. I will explain how it looks: two small sows joined together with only one, perfect head, with its two ears, and at the entrance of the neck, the two other ears together. The body, only one, with its two legs on one side, and on the other side the two other ones. Where the body ends, the abdomens deviate and each have their little thighs on each side and their two tiny paws. In the middle of the abdomen is only one tract through which they both were nourished.[35]

The description given by Paula Montenegro is more detailed than many other letters accompanying monsters shipped to the museum. She seems to have been an aficionada of natural history who wanted to contribute to the museum, much like the other donors mentioned above. Often, descriptions accompanying shipments of monsters were brief and superficial, perhaps revealing that the dispatcher was not all that familiar with the naturalist ways of preserving and describing specimens. If the descriptions accompanying monsters into the museum were more comprehensive, they would normally have been written by a local authority connected to the case due to his profession, such as a surgeon, a priest, a mayor, or any other high official, but rarely by their wives!

In the letter, Montenegro further described how the whole village had looked at the aborted pig, and how she had obtained it and now wanted to embalm it, but that she did not know of any competent person in the village able to perform the embalming. She proposed that the piglet could be sent to Madrid by postal service if it pleased the director. However, she feared that it would putrify, implying that the director should arrange for speedier shipment. Eight days later, a reply was sent to Paula Montenegro stating that the director did not accept the offer—precisely because it was assumed by then to be rotten, due to the fact that she had not been able to find someone in the village to conserve it.

Based on how other monsters were included and described by the donors, it seems unlikely that the rejection solely depended on its possible putrid state. This was a common problem with all specimens to be shipped, and the director could probably have arranged for it to be sent earlier if he wanted. Monsters, although not as rare as in earlier times, were far from everyday occurrences. Such rarities were still considered necessary and valuable objects not only at the Madrid cabinet but also generally among European collectors. One could wonder if other reasons also affected the director's decision to turn down Montenegro's offer. The basis of the rejection actually becomes much clearer when the rest of the letter is taken into consideration. After having offered the piglet to the museum, Montenegro changed the subject, describing how her family had been sent to the village two years earlier due to her husband's military career. They lived very happily there, she wrote. However, there was one major problem. Her husband, a military commander, had not been given the promotion he deserved: "They have forgotten to give him the degree of lieutenant colonel, this being an obvious thing since he has been serving for fifty years." And, she continued, "since no one is doing anything about it, as time goes by, it will all be forgotten."[36] This can be seen as quite a strong intimation from Paula Montenegro to museum director Dávila, urging him to use his position within the royal institutional sphere to influence the right people so that her husband would finally get the promotion he supposedly deserved. The offer seems to have been a somewhat odd combination: by donating a presumably valuable monster she would expect a personal benefit in return. She literally tried to trade a monster for a promotion. The connections between influential persons in important positions at the various royal institutions were tight and the boundaries between the institutions fluid; for instance, many naval and military officers serving the crown, both overseas and on the peninsula, were among the most important contributors to the museum collections. Hypothetically, it was not an impossible exchange, and the director could probably have used his influence to help doña Paula. However, apparently Dávila did not want to use his ascendancy and authority as museum director to do her such a favor.

The rejection of the piglet seems to have had as much to do with the nature of the letter—specifically, the donor's intentions—as with the actual, concrete qualities of the piglet. One could wonder why Montenegro spoke on behalf of her husband. Had he possibly fallen into disgrace with his superiors and could not present his case by himself? This may have been a contributing reason for the letter. In order to understand the rejection, though, issues of gender may be relevant. As a woman, Montenegro was not included in the societies of learned naturalists, nor could she have been a member of the academies—even though she could have been participating in some of the more informal social networks surrounding the museum and connected to science and natural history. The direct participation of women in the Enlightenment sphere of reform and progress was a controversial topic at the time, for instance in the Sociedades Economicas de los Amigos del Pais.[37] These societies were established on private initiatives in all major population centers throughout Spain from the early 1760s onward in order to stimulate the intellectual and economic development of the country by contributing to the progress of industry, agriculture, professions, sciences, and art.[38] In 1787 the presence of women was accepted for the first time, after having been intensely discussed for a lengthy period in the Sociedad de los Amigos of Madrid.[39] In 1783, when Montenegro offered the piglet, women were still not formally included in any of these societies—yet one can imagine how several women were situating themselves at the borders of the male-dominated spheres of science, reform, and progress, trying to negotiate their way in. Paula Montenegro may have been one of these; her description of the piglet reveals that she must have possessed some knowledge of natural history, and the monstrous piglet may have represented an application for her participation in natural history. But deprived of possibilities to formally enter the male-controlled arena, she most likely lacked the relevant knowledge about how to act within this realm.

The intention of using the monster to get her husband promoted was not at all in line with how monsters (or any other object for that matter) ideally should enter the museum, and hence the offer had to be refused, even though the monster itself could have been a valuable contribution to the collection. Montenegro's rather blatant attempt to use the director as an intermediary probably made the offer unacceptable. Not only was she a woman and as such not considered a participant, but neither did she possess the proper savoir faire held by other naturalists and successful museum contributors. Nevertheless, she did try to enter this sphere to the advantage of her husband, and maybe also to promote her own participation as an aficionada, but without any success. One was not supposed to ask so directly for favors, and in the way it was presented this was an unheard-of request; the social implications of a potential acceptance would be too high a price for the director to pay.

### "A Monster That a Sow Gave Birth to, If It Is of Any Interest"

The last monstrous pig of record sent to the museum arrived in November 1798 from don Judas Tadeo Salaura, a clergyman in the village of Guadalix in the diocese of Toledo. In the accompanying letter, this piglet was described as "a monster that a sow gave birth to, if it is of any interest."[40] No response was sent to Salaura, and no note or memorandum about the pig from the museum staff has been kept. Salaura had a lower status within the church than did the archpriest in Úbeda, the donator of the big pig discussed above, and thus was a less prominent contributor to the museum. This might explain why this monstrous specimen apparently received little attention at the cabinet.

However, another and probably more important reason why this piglet never took on an important function was the changing status of the monster collection. Dávila, the first museum director, died in 1786—and the importance given to monsters at the museum depended hugely on his interests. With the arrival of a new management team, a "cleanup" of the collections was initiated in the 1790s. The new director, Eugenio Izquierdo, and vice director, José Clavijo y Fajardo, wanted to rid the museum exhibitions of what they saw as an excessive entertainment factor. Instead, they wanted the museum to be more scientifically oriented.[41] Beginning in the 1790s there was a decrease in the number of documents about monsters at the Royal Cabinet: people stopped sending monsters to the collections, and the museum staff gradually seemed to pay less attention to them. Such marginalization can be seen in relation to the interests of the different directors at the museum, but even more in terms of the general, increasing institutional specialization in contemporary Europe and the move toward teratology as a subdiscipline within medicine. In Spain, Carlos IV established the Real Colegio-Escuela de Veterinaria de Madrid (the Royal Veterinary School of Madrid) in 1792. During the final decade of the eighteenth century and the early nineteenth century, animal monsters became objects of interest at the veterinary school, with a corresponding decline in interest at the Royal Cabinet. To illustrate this, yet another monstrous pig can be mentioned. In 1807, a notice in *Gaceta de Madrid*, the royal official newsletter, announced that an etching of a monster piglet could be purchased at the Escribano bookshop in Madrid. The etching represented "a monster with the head of an elephant, which a sow gave stillbirth to, after having given birth to ten live piglets, at the royal site of San Ildefonso the day March 31, 1807," and it was also announced that the monster was "put on display in the cabinet of the Real Escuela Veterinaria of this court."[42] So, even though the monsters were considered less interesting and relevant at the museum, they did not at all disappear from the public scene as display objects. Rather, they were moved into a more specialized institution: they were put on display in the cabinet at the veterinary school. Thus monsters gradually seemed to lose their didactic and scientific significance at the cabinet, and instead were incorporated by veterinary medicine.

But this incorporation is not the entire story. Due to a constant lack of space at the cabinet, the museum did not have sufficient room to store objects. The only space available was in the attic, which had rooms used for lodging the employees. The rooms occupied by the management were often used to store objects that could not be housed in the display halls, but these were jam-packed decades before the monsters started to lose significance.[43] Hence the monsters were still kept on display, even though the written records testify to less attention being paid to the monsters. In 1785 the construction of a grand new scientific academy—Academia de Ciencias—in el Paseo del Prado had started, and in this planned academy the cabinet would have a central position with abundant spacious halls for display, storage, and preparation. But when the building was finally completed around 1818, the king decided that it was to be used as a *pinacoteca*, an art museum, instead of a scientific academy, clearly implying that art, and not the natural sciences, was to be the preferred means for representing the Spanish Empire.[44]

## Pig Monsters and Museum Practices

The archival documents reveal that the pig monsters were represented in many different ways and that they were incorporated into a multitude of museum practices. All the pig monsters were flexible objects that could be shaped by describing them according to different textual practices, practices that transformed the monsters into something they would not have been without that specific document. Monsters and monstrosities were used, explained, negotiated, and represented vis-à-vis different audiences that became connected to the museum, toward different ends.

The piglets donated by Gaspar Soler and the Duke of Híjar gained importance by being offered and accepted as gifts to the cabinet—and by this the monster became an important means of establishing and situating the museum within the sphere of reform and progress in eighteenth-century Spain. The piglets materialized important relations within a socio-institutional context: between donors within the elite, museum management, high state officials, king, and court. These relations took place within and sustained a complex network of science, reform, and royal patronage where the institutional boundaries were fluid. But it was not for everyone to enter this sphere, and for some the boundaries were impossible to cross. The case of the military wife Paula Montenegro demonstrates this. Yet another piglet, "the rarest character among the monsters," was represented according to a textual genre known to the eighteenth-century naturalist, and functioned as an interesting object to be reflected on within a learned milieu of aficionados and privileged guests. Within such a socio-scientific context the descriptions were focused on characteristics of the pig. In this case its provenance and its donor were not as important as the monster itself.

The clergy was an important contributor to the museum, and two of the pigs were donated by clergymen. A monstrously big pig was donated by the archpriest Francisco de Guya, and in the museum it probably served, due to its mere grandeur, to arouse patriotism by representing one of the many marvels of the Spanish Empire. On the other hand, the big pig was also implicitly incorporated into a normalizing practice by making manifest understandings of the normal pig. However, during the course of some decades the big pig's size lost its significance. The last pig discussed was a small one, and one that did not seem to gain much importance, due to the changing status of the monster collection. It seems as if the donor, Salaura, a clergyman from Guadalix, did not even receive a response to his offer. Thus, both of the last pigs represent the reduced significance given to monsters at the museum at the turn of the century.

The archival remnants of the pigs, then, tell us about a multitude of museum practices. But they can also reveal something about how representations and understanding of monsters changed toward the end of the eighteenth century at the Royal Cabinet in Madrid—of how animal monsters gradually became objects of a specialized subdiscipline within medicine, and lost significance as objects for display and study in the cabinet. The contributors to the museum stopped sending in monsters, and the museum employees gradually lost interest in them—monsters no longer constituted the valuable contribution to the museum that they once did. The study of the pig monsters in the Madrid cabinet adds to the picture of the eighteenth century as a period of rationalization, domestication, and normalization of the monstrous.

The pigs were transformed, through their circulation between people, by being represented in texts and by being discussed and displayed, by being accepted or rejected. To see all these incidents and actions as forming part of a progressive movement toward nineteenth-century teratology, however, would be too narrow a focus. It is doubtful that, for instance, Gaspar Soler, the Duke of Híjar, or Paula Montenegro had scientific progress as a chief motivation when they offered the pigs to the cabinet. What they wanted, as members of the societal elite, was to enhance their own position and become connected to a learned milieu in the capital city. The clergymen de Guya and Salaura acted on behalf of the church, by donating presumably valuable monsters; one motivation was probably to strengthen the position of the church vis-à-vis the king. Even if the actions of all the actors that have been discussed in this essay in some way contributed to a process culminating in the later teratology and in the *Entzauberung* of the natural world, it would make the analysis poorer if the historical details of their actions were left out. The history of monsters consists of a composite where scientific practices, personal needs, institutional norms, political strategies, and accidental incidents blend together.

## NOTES

1. This is an estimate based on letters and other documents from the Royal Cabinet in the period 1773–1800 as registered in Calatayud's *Catálogo de Documentos del Real Gabinete de Historia Natural, 1752–1786*; and *Catálogo Crítico de los Documentos del Real Gabinete de Historia Natural, 1787–1815*.

2. *Entzauberung* was first coined by the German sociologist Max Weber and particularly elaborated in his early twentieth-century work *The Protestant Ethic and the Spirit of Capitalism*.

3. The establishing of teratology as a discipline is closely connected with the work of Étienne Geoffroy Saint-Hilaire and his son Isidore Geoffroy, culminating in the grand *Traité de Tératologie* (1832–37).

4. See Daston and Park, *Wonders and the Order of Nature*; Moscoso, "Monsters as Evidence"; Hagner, "Enlightened Monsters"; Lafuente and Moscoso, *Monstruos y seres imaginarios en la Biblioteca Nacional*; Knoppers and Landes, *Monstrous Bodies/Political Monstrosities in Early Modern Europe*; Maerker, "Tale of the Hermaphrodite Monkey."

5. Hagner, "Enlightened Monsters," 178.

6. Curran and Graille quoted in Knoppers and Landes, "Introduction," 5.

7. Daston and Park, *Wonders and the Order of Nature*, 17.

8. Ibid., 360.

9. Knoppers and Landes, "Introduction," 11.

10. See Barras y de Aragón, "Cerdo Monstruoso."

11. On the history of El Real Gabinete de Historia Natural, see Villena et al., *El gabinete perdido*; Calatayud, *Pedro Franco Dávila*; Barreiro, *El Museo Nacional de Ciencias Naturales*; Pimentel, *Testigos del mundo*; Lafuente, *Guía del Madrid científico*; González-Bueno, "El Real Gabinete de Historia Natural"; Schulz, "Spanish Science and Enlightenment Expeditions." Among the advocates for establishing a museum of natural history were the physician and botanist José Celestino Mutis, the Augustine clergyman Enrique Flórez, and the minister of economy and finance Pedro Rodríguez Count of Campomanes. See Calatayud, *Pedro Franco Dávila*, 43, 65, 81; and Wulff, *Las esencias patrias*, 67. For an overview of eighteenth-century Spanish history and Bourbon reforms, see Ruiz Torres, "Reformismo e Ilustración."

12. Calatayud, *Pedro Franco Dávila*, 104, 111.

13. In 1776 an instruction written by Dávila was sent to representatives of the crown in all Spanish territories, with the pronounced purpose to augment the collections of the Royal Cabinet of Natural History. The instruction formed part of a larger initiative organized by the Spanish crown to have local administrators in the Americas and the Philippines collect natural history specimens and send them to Spain. See Calatayud, *Pedro Franco Dávila*; De Vos, "The Rare, the Singular, and the Extraordinary"; Villena et al., *El gabinete perdido*. The image of the monstrous pig (fig. 1.1) was probably shipped to Spain due to this order.

14. Dávila was a member of several academies, such as the Prussian Imperial Academy of Science; the Imperial Academy in St. Petersburg; the Real Academia de Ciencias in Madrid; the Reales Academias de Buenas Letras in Madrid, Seville, and Biscay; and the Real Academia de Historia de Madrid. In 1776, he was accepted as a fellow of the Royal Society in London after being recommended by Comte de Buffon. See Calatayud, *Pedro Franco Dávila*, 126. Dávila also had contact with Linnaeus in Sweden. As a private collector in Paris, he hired Romé de l'Isle, a well-known naturalist at the time, to make a catalog of his private collection. The three volumes were published in 1767, with a foreword and approval written by Michel Adanson. See Pimentel, *Testigos del mundo*, 176. In 1777, Eugenio Izquierdo was appointed vice director and professor of natural history at the cabinet, and he was promptly sent out to travel, in order to study and visit cabinets in Europe, which reflects the importance given to the participation in European networks of natural history and science at the Royal Cabinet. See Calatayud, *Eugenio Izquierdo de Rivera y Lazaún*.

15. In 1790, another Spanish royal cabinet of natural history was opened in Nueva España by Joseph Longinos Martínez, who participated in a royal scientific expedition. The opening was announced in the royal newsletter, *Gaceta de Madrid*, no. 86, October 26, 1790.

16. See Medina, *Espejo de sombras*; Lafuente and Igea, "La construcción de un espacio público para la ciencia."

17. Ref. 211 in Calatayud, *Catálogo de Documentos del Real Gabinete de Historia Natural, 1752–1786*.

18. Ref. 247 in ibid. The mines in Almadén had been royal property since 1645. In 1777, King Carlos III founded the Academia de Minería y Geografía Subterranea in Almadén.

19. Ref. 947 in ibid.

20. Ref. 836 in ibid. The name of the hospital is not mentioned, but it was likely the Hospital General de Madrid. See Lafuente, *Guía del Madrid científico*.

21. See Spary, *Utopia's Garden*, 36; and Schulz, "Spanish Science and Enlightenment Expeditions," 193, on patronage relations in France and Spain.

22. See Moscoso, "Monsters as Evidence."

23. Ref. 574 in Calatayud, *Catálogo de Documentos del Real Gabinete de Historia Natural, 1752–1786*.

24. The image shown above (fig. 1.1) made by a Spanish official in the Philippines represents an apparently similar monstrous specimen. In the collection Dávila brought from Paris, he had a monstrous pig with similar characteristics: a one-eyed piglet with three incisors in the upper jaw. In the catalog of Dávila's Paris collection the description of this monster is given, and it is similar to the one about "the rarest character," enhancing only the most singular features. Dávila and de l'Isle, *Catalogue systématique et raisonné des curiosités de la nature et de l'art*, 499.

25. Moscoso, "Monsters as Evidence," 355.

26. Ibid., 358.

27. Hagner, "Enlightened Monsters," 177.

28. Curran, "Afterword," 231.

29. Refs. 384 and 842 in Calatayud, *Catálogo de Documentos del Real Gabinete de Historia Natural, 1752–1786*.

30. Hagner, "Enlightened Monsters," 194; Moscoso, "Monsters as Evidence."

31. Ref. 384 in Calatayud, *Catálogo de Documentos del Real Gabinete de Historia Natural, 1752–1786*.

32. Colina, *El folklore en la obra de José Pereda*, 96.

33. Ref. 554 in Calatayud, *Catálogo de Documentos del Real Gabinete de Historia Natural, 1752–1786*.

34. Mieg, *Paseo por el gabinete de historia natural de Madrid*, 104.

35. Ref. 718 in Calatayud, *Catálogo de Documentos del Real Gabinete de Historia Natural, 1752–1786*.

36. Ibid.

37. Castells, Espigado, and Cruz, *Heroínas y patriotas*, 38.

38. The founder of these *sociedades*, Pedro Rodríguez, Count of Campomanes, was one of the voices that during the 1760s recommended that Carlos III establish a cabinet of natural history in Madrid. See Calatayud, *Pedro Franco Dávila*, 65.

39. Castells, Espigado, and Cruz, *Heroínas y patriotas*, 38.

40. Ref. 337 in Calatayud, *Catálogo Crítico de los Documentos del Real Gabinete de Historia Natural, 1787–1815*.

41. Barreiro, *El Museo Nacional de Ciencias Naturales*, 82.

42. *Gaceta de Madrid*, no. 39, May 5, 1807.

43. Ref. 366 in Calatayud, *Catálogo de Documentos del Real Gabinete de Historia Natural, 1752–1786.*

44. Lafuente, "La colina de las ciencias."

## BIBLIOGRAPHY

Barras y de Aragón, Francisco de las. "Cerdo Monstruoso." *Boletín de la Real Sociedad Española de Historia Natural* 46, nos. 9/10 (1948): 762–63.

Barreiro, Augustín J. *El Museo Nacional de Ciencias Naturales, 1771–1935.* Aranjuez: Doce Calles, 1992.

Brú y Ramon, Juan Bautista. *Coleccion de laminas que representan los animales y monstruos del Real Gabinete de Historia Natural de Madrid, con una descripcion individual de cada uno.* Madrid: Andres de Sotos, 1784–86.

Calatayud Arinero, María. *Catálogo Crítico de los Documentos del Real Gabinete de Historia Natural, 1787–1815.* Madrid: Consejo Superior de Investigaciones Científicas, 2000.

———. *Catálogo de Documentos del Real Gabinete de Historia Natural, 1752–1786.* Madrid: Consejo Superior de Investigaciones Científicas, 1987.

———. *Eugenio Izquierdo de Rivera y Lazaún, 1745–1813: Científico y político en la sombra.* Madrid: Museo Nacional de Ciencias Naturales, Consejo Superior de Investigaciones Científicas, 2009.

———. *Pedro Franco Dávila: Primer director del Real Gabinete de Historia Natural fundado por Carlos III.* Madrid: Consejo Superior de Investigaciones Científicas, Museo Nacional de Ciencias Naturales, 1988.

Castells Oliván, Irene, Gloria Espigado Tocino, and María Cruz Romeo. "Heroínas para la patria, madres para la nación: Mujeres en pie de guerra." In *Heroínas y patriotas: Mujeres de 1808*, edited by Irene Castells Oliván, Gloria Espigado Tocino, and María Cruz Ro, 15–54. Madrid: Cátedra, 2009.

———, eds. *Heroínas y patriotas: Mujeres de 1808.* Madrid: Cátedra, 2009.

Colina de Rodríguez, Luz. *El folklore en la obra de José Pereda.* Santander: Institucíon cultural de Cantabria, 1987.

Curran, Andrew. "Afterword: Anatomical Readings in the Early Modern Era." In *Monstrous Bodies/Political Monstrosities in Early Modern Europe*, edited by Laura L. Knoppers and Joan B. Landes, 227–45. Ithaca: Cornell University Press, 2004.

Daston, Lorraine, and Katharine Park. *Wonders and the Order of Nature, 1150–1750.* New York: Zone Books, 2001 [1998].

Dávila, Pedro F., and Romé de l'Isle. *Catalogue systématique et raisonné des curiosités de la nature et de l'art: Qui composent le cabinet de m. Davila.* Paris: Chez Briasson, 1767.

De Vos, Paula. "The Rare, the Singular, and the Extraordinary: Natural History and the Collection of Curiosities in the Spanish Empire." In *Science in the Spanish and Portuguese Empires, 1500–1800*, edited by Daniela Bleichmar, Paula de Vos, Kristin Huffine, and Kevin Sheehan, 271–89. Stanford: Stanford University Press, 2009.

Geoffroy, Isidore Saint-Hilaire. *Histoire générale et particulière des anomalies de l'organisation chez l'homme et les animaux.* Paris: J.-B. Baillière, 1832–37.

González-Bueno, Antonio. "El Real Gabinete de Historia Natural." In *Madrid, ciencia y corte*, edited by Antonio Lafuente and Javier Moscoso, 247–51. Universidad de Alcalá: Consejería de Educación y Cultura, Consejo Superior de Investigaciones Científicas, 1999.

Hagner, Michael. "Enlightened Monsters." In *The Sciences in Enlightened Europe*, edited by William Clark, Jan Golinski, and Simon Schaffer, 175–217. Chicago: University of Chicago Press, 1999.

Knoppers, Laura L., and Joan B. Landes. "Introduction." In *Monstrous Bodies/Political Monstrosities in Early Modern Europe*, edited by Laura L. Knoppers and Joan B. Landes, 1–22. Ithaca: Cornell University Press, 2004.

———, eds. *Monstrous Bodies/Political Monstrosities in Early Modern Europe*. Ithaca: Cornell University Press, 2004.

Lafuente, Antonio. "La colina de las ciencias." In *Madrid, ciencia y corte*, edited by Antonio Lafuente and Javier Moscoso, 229–37. Alcalá: Consejería de Educación y Cultura, Consejo Superior de Investigaciones Científicas, 1999.

———. *Guía del Madrid científico: Ciencia y corte*. Madrid: Doce Calles, Comunidad de Madrid, Consejería de Educación y Cultura, Dirección General de Investigación, CSIC, 1998.

Lafuente, Antonio, and Javier Moscoso. *Monstruos y seres imaginarios en la Biblioteca Nacional*. Madrid: Ministerio de Educación y Cultura, Biblioteca Nacional, 2000.

Lafuente, Antonio, and Juan Pimentel Igea. "La construcción de un espacio público para la ciencia: Escrituras y escenarios en la ilustración española." In *Historia de la ciencia y de la técnica en la corona de Castilla*, vol. 4, edited by José Luis Reig, 111–56. Salamanca: Junta de Castilla y León, 2002.

Maerker, Anna. "The Tale of the Hermaphrodite Monkey: Classification, State Interests, and Natural Historical Expertise Between Museum and Court, 1791–4." *British Journal for the History of Science* 39, no. 1 (2006): 29–47.

Medina Domínguez, Alberto. *Espejo de sombras: Sujeto y multitude en la España del siglo XVIII*. Madrid: Marcial Pons, 2009.

Mieg, Juan. *Paseo por el gabinete de historia natural de Madrid, ó, Descripción sucinta de los principales objetos de zoología que ofrecen las salas de esta interesante colección*. Valladolid: Editorial Maxtor, 2009 [1818].

Moscoso, Javier. "Monsters as Evidence: The Uses of the Abnormal Body During the Early Eighteenth Century." *Journal of the History of Biology* 31, no. 3 (1998): 355–82.

Pimentel Igea, Juan. *Testigos del mundo*. Madrid: Marcial Pons, Ediciones de Historia, 2003.

Ruiz Torres, Pedro. "Reformismo e Ilustración: Volumen 5." In *Historia de España*, edited by Josep Fontana and Ramón Villares. Barcelona: Crítica; Madrid: Marcial Pons, 2007.

Schulz, Andrew. "Spaces of Enlightenment: Art, Science, and Empire in Eighteenth-Century Spain." In *Spain in the Age of Exploration, 1492–1819*, edited by Chiyo Ishikawa, 189–227. Lincoln: University of Nebraska Press, 2004.

Spary, Emma C. *Utopia's Garden: French Natural History from Old Regime to Revolution*. Chicago: University of Chicago Press, 2000.

Villena, Miguel, Javier Ignacio Sánchez Almazán, Jesús Muñoz Fernández, and Francisco Yagüe Sánchez. *El gabinete perdido: Pedro Franco Dávila y la historia natural del siglo de la luces: Un recorrido por la ciencia de la ilustración a través de la "producciones

marinas" del Real Gabinete (1745–1815). Madrid: Consejo Superior de Investigaciones Científicas, 2009.

Weber, Max. *The Protestant Ethic and the Spirit of Capitalism*. Translated by Talcott Parsons. London: Allen and Unwin, 1971.

Wulff Alonso, Fernando. *Las esencias patrias: Historiografía e historia antigua en la construcción de la identidad española (siglos XVI–XX)*. Barcelona: Crítica, 2003.

**2**

# The Frames of Specimens: Glass Cases in Bergen Museum Around 1900

*Brita Brenna*

Glass is, in general, the enemy of secrets.
—WALTER BENJAMIN

In his 1914 treatise *Glasarchitektur*, Paul Scheerbart announces the arrival of "a culture of glass." In this new glass milieu, he claims, humanity will become completely transformed. So will nature: "The whole of nature will in all regions of culture appear to us in a quite different light, after the introduction of glass architecture."[1] Scheerbart's theme was glass as a material for new buildings, private as well as public. The material would be a means of reforming the sensibility of people, the nature of society, and the perception of nature. In this chapter my ambitions are more modest. Focusing on glass cases in museums, I question how they frame and allow us to know nature. What is the nature of the "museum nature" that glass cases enclose and make visible? "Museum nature" is here used to signify "the practices of collecting, preservation, and displaying certain things—animals, plants, fossils and rocks—and the conceptual and exhibitionary frameworks in which they are set."[2] Museum nature is a term for what the museum produces: a very specific kind of nature, manufactured through processes of care, curiosity, disciplined work, and pedagogic ambitions. I will use the following pages to query what role glass and the glass case play in constructions of museum nature and, in particular, constructions of animals.

A recent visit to American zoos made me urgently aware of the persistence of glass as a medium for experiencing animals. To help us look at animals in their carefully orchestrated fields, the zoos provide shelters with panoramic panes, offering an optimal viewpoint and a medium through which to look. Live animals are offered to the viewer as though they were stuffed animals in a museum diorama, in a glass case. Pet shops offer the same experience, presenting animals staged in total visibility but behind glass. "Man becomes aware of himself returning the look [of the animal]," John Berger wrote.[3] Indeed, glass allows us to be in visual communion with the animal. But what does man become aware of when returning the look through a glass pane? Glass allows communication but prevents anything other than scopic engagement. The whole of nature will look different through glass, Scheerbart contended, and so my concern is to interrogate what glass as a medium does to the experience of animal nature.

My discussion focuses on Bergen Museum and its exhibitionary practices in the decades around 1900. Established in 1825, Bergen Museum was the first purpose-built museum in Norway.[4] At this time, the city of Bergen was a thriving merchant town, and its residents invested much prestige and money into the museum project. The well-to-do and well-reputed citizens were members of the museum society and the board. In fact, board members were responsible for a substantial part of the museum's research, collection, and conservation activities until the end of the nineteenth century. The collection policy was expansive, the ambitions became universal, and the collection mushroomed during the nineteenth century as the museum assembled natural, historical, and ethnographical objects from Norway as well as specimens from overseas. The monumental stone building, inaugurated in 1867, was expanded with two new wings in 1898, fulfilling a long overdue extension of the premises.[5] Archival materials held at the museum allow us to question how, why, and what museum nature was collected, cared for, and displayed. In the last two decades of the nineteenth century, the interest in how to collect, store, and exhibit nature was prevalent in the natural history department, which will serve as my focus. This period also coincided with lively international debates about museum education, museum science, and museum policy—debates that reverberated around the organization and new architecture of Bergen Museum.[6]

A pertinent question is whether it is possible to put the glass case, a universal exhibition tool, up for questioning based on material from a regional museum in Norway in the decades around 1900. Certainly, collection, exhibition, and scientific practices were, as I will show, going through a period of intense standardization internationally. This makes a local and minor museum interesting as an object of analysis, highlighting both local specificities and how they might partake in framing the "universal." However, the universality and prevalence of glass cases in museums make it necessary to investigate the tension between the particular glass cases and the emblematic status of the glass case in our perception of museums. The questions are certainly entangled: How do glass cases construct museum nature in Bergen, and how do glass cases construct "the museum"? Glass cases are ubiquitous and all-important in museum work, yet transparent and rendered invisible; their movement and whereabouts, their construction and work, are difficult to track down in museum archives. Rather than presenting a finished set of answers, then, this chapter attempts to raise questions about what happens when we insist on treating glass cases as important epistemological tools in museum work, in science, and in science education.

The German art critic, philosopher, and writer of futuristic novels Paul Scheerbart (quoted above) wrote his treatise as an inaugurational text announcing a culture of glass. Glass culture was for him both contemporary and a fact of the future. However, Scheerbart wrote at the end of a century that had witnessed an incessant occupation with glass. Glass had been an important tool in forging Victorian culture, as Isobel Armstrong demonstrates in *Victorian Glassworlds* (2008). She reveals a glass culture that has little to do with twentieth-century modernism, and everything to do with nineteenth-century modernity.[7] Here her argument serves as an impetus for looking at glass cases not as functionalist vitrines in nineteenth-century museums, born ahead of time, but as firmly grounded in nineteenth-century culture. Glass cases were ubiquitous tools for exhibitionary purposes, sprawling private homes, luxury shops, and department stores, as well as museums and public expositions. Behind windowpanes, which were constantly growing larger, animals were exhibited as amusements or object lessons, as trophies or scientific specimens.

## A Knowledge Culture of Glass

Glass cases fulfill multifarious functions, the most obvious being to make visible and to make untouchable: simultaneous display and protection. Where can we trace their history? One direction is genealogical. In her article "Vom Einräumen der Erkenntnis," historian of science Anke te Heesen accounts for the different furniture forms that have left their traces on contemporary scientific cupboards and cases. She deals with the kind of furniture that is so inconspicuous that we seldom if ever see it as furniture, and even less as phenomena to be contemplated. Heesen addresses scientific furniture in different forms.

For her, storage is the obvious common denominator. "All knowledge requires a container," she claims, outlining a short history of the storage of knowledge.[8] The chest and the case for relics were the prevalent forms in the middle ages. Gradually, the buffet developed into a form that contained both open shelves and closed containers, thus both exposing and enclosing the contents. From the Renaissance we also have the *Kunstschrank*, the furniture for the treasures of kings and the wealthy, where the different realms of the world were tucked away in numerous drawers with the most exquisite ornamentation.[9]

These examples point to the importance of containers for keeping things safe—an obvious point, but still useful to bear in mind when we encounter glass cases. Whether the objects in early collections were on shelves or on tables in a room, in the drawers of a *Kunstschrank* or inside a closed cabinet, they would need a person to attend if somebody wanted to look at them. During the eighteenth century, cabinet doors were fitted with glass plates, thus making it possible for a person to inspect the contents without any-one intervening: "Here there was no more talk of hands, but of eyes."[10] This, Heesen writes, set the standard for the new creation developing during the eighteenth century—the publicly accessible museum.[11] Glass panes made it possible to double the space of a collection, making an inside and an outside. They produced one space sealed off from the public, accessible only to the custodian or the owner, and another for the visitor who could now enter the museum as part of an anonymous public, who did not have to relate person-ally to the caretaker of a collection.

As seen in manuals, prints, and paintings of collections and museums from the eighteenth century onward, nature could now be inspected through glass while being safely out of the viewer's reach. The importance of glass is obvious, for example, in Charles Wilson Peale's painting of himself lifting the curtain to unveil the rows of birds in glass boxes in his Philadelphia museum, and in prints of William Bullock's collections in London and Liverpool from the first decades of the nineteenth century. Birds, animal fossils, and insects were to be seen behind glass in Peale's collection. In Bullock's London Museum, there appeared glass cases that were more than two meters high, and with other cases still on top of them. The glass cases held ethnographic material, birds, and preserved animals, and the glass must have been of sufficient quality to see the contents adequately without needing to open the doors.[12] Both men operated at the beginning of the nineteenth century, at a time when the "museum move-ment" was in its infancy. Susan Pearce has claimed that Bullock was in fact the first to combine a Linnaean-based arrangement of animals with a naturalistic setting for them—and that "the style of the displays that he mounted can fairly be claimed to have invented a new visual language of objects, one which museums across the globe have, one way or another, been working through ever since."[13] She emphasizes how Bullock exhibited animals in settings without glass around them, a kind of diorama avant la lettre. What the picture points to is that animals did not have to be in glass cases, that there were other ways of exhibiting them, and thus that the choice of using glass cases was conscious and definite. And certainly, some specimens continued to be exhibited outside glass cases, such as bulky stuffed animals, trophies, and large skeletons.

During the nineteenth century, glass became an indispensable tool for making museum nature. To paraphrase Scheerbart, who awaited a new sensitivity to nature developing through the glass and iron buildings of the twentieth century, nature would now be seen in a completely new light. Indeed, this was exactly what happened during the nineteenth century once glass became cheaper, new methods for production were introduced, and glass in all facets became an important tool for forging knowledge cultures, from the glass in the microscope to the glass case.[14] The culture of glass refined and magnified ways of seeing nature through glass.

Glass was an indispensable tool for what Tony Bennett terms "the exhibitionary complex," the ensemble of institutions utilized as vehicles for turning crowds into a docile public. They were sites for the entanglement of official and popular culture: arcades, department stores, national and international exhibitions, and the panoramas and dioramas of museums.[15] Exhibitionary techniques and architectural forms evidence the exchange between these different institutions. But animals and nature also moved between them. Bullock and Peale both operated at the intersection of popular spectacle and serious education. Later in the century, we see the art of taxidermy in different locations, with different agendas and publics. One well-known example is Walter Potter's tableaux, with large groups of guinea pigs, rats, squirrels, kittens, and rabbits performing human tasks.[16] They are carefully set within glass cases that encourage visitors to view them as total ensembles. As Michelle Henning has shown, various practices of anthropomorphic taxidermy existed concurrently, and they differed again from museum taxidermy.[17]

## The Nature of Exchange: Procuring Specimens

When the young curator of Bergen Museum, Fridtjof Nansen, arrived in Hamburg in February 1886, he visited the zoological garden, the art museum, and the opera on his first day. In a letter to his peer, the physician Daniel Danielsen, Nansen complained that Sunday was a bad day to arrive, as to start by resting was against all principles, except for professional loafers.[18] This may indicate both his wish to show how thrifty he was in a letter to the director of the board of the museum, and that hard work and scientific study was the order of the day for employees at the museum during these years. The promising student Nansen had been hired as a curator at the museum in 1882, at the age of twenty-one.[19] Although there were just two curators in the department, there are obvious signs of the ambition to make the museum into a modern institution for research and display. Several of the board members were highly esteemed scientists. The aspirations of Bergen Museum suffused the rhetoric of the young scientist Nansen. The next day he carried out more serious work, visiting the Naturhistorische Museum in Hamburg. He was shown around by the welcoming director, Heinrich Pagenstecher, but was disappointed by what he saw. The museum was a mere "transport horse," he claimed, describing his hometown museum in Bergen as a "race horse" in comparison. However, he would contend, "I still received a whole lot of useful advice."[20]

Indeed, Nansen compared and consulted; this was how museums kept up to date, and it was how Bergen Museum developed strategies for exhibition, conservation, and collecting. The Naturhistorische Museum in Hamburg had knowledge and expertise that were needed in Bergen. One of the striking qualities of museum work in the nineteenth century was the intensive exchange of information. Museums were few, and for museum employees, the best sources of useful information were just as often found in museums across national and continental borders as they were in nearby cities.[21] This was, of course, a quandary for natural history museums; audiences were local, while the ambitions and networks of the employees were international. The same can be said of many scientific institutions, but this bifurcation was perhaps particularly strong in natural history museums because, having public instruction as one of their agendas, they had to relate to their public while also taking part in international research.

Knowledge and work practices were matters of exchange, but so were zoological specimens. In his letter, Nansen gave detailed accounts of his meetings with dealers in natural objects, and of what specimens he acquired for the museum. He was most enthusiastic about the exchange deal he set up with the director in Hamburg. From Bergen they could send off a whale skeleton; from Hamburg they would receive a pickled gorilla, a pickled chimpanzee, and a manatee skeleton. And still they would have the credit of half the whale's

FIG. 2.1 A beaked whale outside Bergen Museum arriving from Nordfjord in 1901. Picture Collection, University of Bergen Library.

worth. Bergen Museum, like other museums, relied heavily on exchange to develop collections that would engage the public and make good research material. The holding of whales in various states at the museum proved an invaluable asset in the exchange market for natural objects, both with professional dealers and with other museums, and Nansen took advantage of this (fig. 2.1).[22] Nansen's dealings are but a tiny part of the immense traffic in specimens—in animals, minerals, and plants—and this exchange took many forms and involved a motley crew. Within just a couple of months in 1888, we find the zoological department in the museum corresponding with museums and individuals from across the Western world, from Minnesota to Vienna, Tromsø to Weymouth, Washington to Prague. The list of individuals involved includes clergymen, museum curators, hunters, and small and large dealers in animal specimens.[23] This was far from exceptional; museum business was international business, and it was indeed a business.

Bergen Museum thrived and acted as a "racehorse" in this period thanks largely to the city's flourishing commercial sea trade and, specifically, Bergen's proximity to commercial whaling activities. As has been thoroughly shown elsewhere, natural history as well as anthropology collections of the nineteenth century were deeply embedded in global economic systems of trade and exploitation—that is, in colonialism.[24] Part of this system involved a race for safeguarding specimens of interest to the museums—the rarer the better. Preserving nature in the museum was seen as a method of saving nature, even though it meant killing the animals in question.[25] In 1893, shipment of a monk seal to Bergen was announced, an animal that was becoming "extremely rare" and could be found only "off the wildest parts of the rocky coasts of Sardinia," according to Direzione del Museo Zoologico dei Vertebrati in Florence.[26] Within these networks, museums were to ensure they all had specimens that could be kept for eternity, and so in this context, killing members of a threatened species was a way of securing for them eternal life.[27]

As with all museums, Bergen Museum had other means of procuring zoological specimens: a taxonomy of acquisition methods lists as possibilities gift, purchase, fieldwork, exchange, and loan.[28] These were all adopted in Bergen, often in hybrid forms. Dealers in natural objects had operated for centuries, but there was a marked increase in the number of participants in networks of exchange and commerce during the nineteenth century. In Bergen, museum staff dealt with professional firms, among others Émile Deyrolle in Paris, J. F. G. Umlauff in Hamburg, and Rowland Ward in London. These well-known firms were dealing in a variety of fields, from scientific instruments to ethnographic objects, from parts of human anatomy to aquariums. That the dealers were not always considered to have thorough knowledge is evidenced in Nansen's letter. He described the persons he dealt with at the premises of Umlauff as "two wise chickens," as he was unsure of whether they had any idea about what a splendid exemplar of a killer whale he was offering them.[29]

Locals were less commercial but not necessarily less professional. The wealthy local furrier, Brandt, for example, would furnish the museum with new specimens from the local fauna, as the mounted species became old and worn.[30] The museum would also provide seafaring men, captains, and others with dredges, glass containers, and alcohol so that they could bring new species home.[31] In addition, locals constantly brought or sent more or less well-preserved specimens to the collection. Lastly, there was a budget for buying, albeit a small one. I have emphasized the exchanges of specimens, not because they were particular to Bergen but because they show the extent to which natural history museums were part of networks that produced museum nature. The museum was only one of the sites where this process took place, and the nature that became visible there was a result of a long chain of negotiations and transformations. When zoological specimens entered the museum, these negotiations continued. In the following section I will not follow the specimens through the different parts of the museum, but rather bring them straight to the question of ordering display.

## Reorganizing Museum Nature

Networks for the collection of scientific specimens were accompanied by networks for standardizing display. Jørgen Brunchorst, hired in 1886 as a curator in the botanical department in Bergen Museum, is of special interest in this regard. He was particularly eager to reform and reorder the museum, and became the first secretary to the natural history department and later the first director of the museum itself. Brunchorst, born in Bergen, had studied plant physiology at various universities in Germany before being appointed curator of the botanical collection. Throughout the 1890s, the museum's activity was fervent, and Brunchorst was taking the lead. The museum opened a biological research station, ran a summer course for schoolteachers, presented a popular lecture series titled "Lectures for Everyman" (which offered 125 lectures in 1894), published the leading Norwegian scientific journal, *Naturen*, opened a botanical garden, and not least remained active in advertising the museum as the foundation of a new university in Norway. Crowning these efforts were the museum's two new wings, inaugurated in 1898.

Ambitions to reform the organization and display of natural specimens must be seen against this backdrop of networked activity. Reorganization of the display was a means of securing the double mission of the museum: to educate the common man, and to be an institution of higher education and research. In this, display cases had important functions to fulfill. They took part in shaping new divisions within the museum, they materialized pedagogic principles, and they helped create variously coded spaces within the museum. The 1894 protocol of the natural history department shows how specimens were moved around to new locations, old glass cases were used for new purposes, some were rebuilt, others moved, and new ones ordered.[32] Throughout the same year, the total cost for "things of glass" was 145 kroner, for alcohol 171, for natural objects 859, and for rebuilding old and buying new furniture 1,323 kroner.[33] Furniture was the single largest expense in the department.

The need for more satisfying storage and exhibition space was an important argument justifying an increase in the budget.[34] "In the coming years," wrote Brunchorst, "it is just as necessary to have money at disposal for a more extensive translocation in the collection, to procure new and rebuild older cupboards and glass cases. When a collection grows as time passes, sooner or later there will occur a moment when it is no longer possible to push the new artifacts between the older ones, when the whole arrangement has to be modified, the glass cases to a large extent rebuilt and changed, if good order should be sustained."[35] In this instance, the museum argued for glass cases as means of maintaining good order, yet glass cases did not as such impose a preconceived order on the displays themselves. The nature of the zoological specimens that were to be exhibited, and the system in which they were to be displayed, would influence the standard the glass cases would help express. If we look at Brunchorst's own writings on the topic, it becomes evident that the refurbishing of the department was closely connected to new, broader visions for museums in general, and to a new role for the zoological specimens held and displayed by those museums in particular.

In 1890, Brunchorst wrote in the museum's yearbook about a study trip to Britain, reflecting in particular on the natural history museum in South Kensington (British Museum, Natural History). He was deeply impressed by what he saw in London, and it is safe to say that this trip was decisive in turning his attention toward the museum as a promising public and pedagogic institution. He admired the glass cases in South Kensington, writing, "They are quite elegant, but also very expensive, as mahogany, glass, and wrought iron are the predominant materials used for desks, as well as for freestanding glass cases and wall cabinets." Systematic displays, separate departments for research collections, instructive and detailed labels, and illustrations in the forms of maps and drawings are all characteristics of the collections that he emphasized with approval. In all the departments that caught his attention there had been "great emphasis on communication of knowledge to the visiting public; with an emphasis on forcing the visitors not simply to satisfy their curiosity, but really to learn something." This was even more so, he claimed, in regard to the "Introductory Collection," which was installed in the hall of the museum: "This collection is an elaborate and comprehensive textbook in 'general zoology' and 'general botany.'" Brunchorst described the collection as a textbook that had been paraphrased onto labels that meticulously described every specific object. The labels explained the specimens, and the specimens served as illustrations for the labels, as he described it. "After a thorough examination of one of these glass cases one has been taught many hours' worth of zoology within less than half an hour," Brunchorst exclaimed. He pointed to the extensive work behind such a collection for instruction, but had to admit that it would not be within reach for most museums to even think about such a comprehensive design. Still, he prophesied that establishing teaching collections analogous to this one would become a necessity for all museums that wanted to stay abreast of the times.[36]

Throughout the yearbook entry, Brunchorst stressed that the objects needed to be instructive for visitors and accessible for scientists. He found the labeling to be of particular importance for the public, but he was also impressed by different kinds of models in wax, glass, or plaster. The most instructive of all the departments was the series of British birds placed in their natural habitats, accompanied by true-to-nature copies of the relevant plants. The copies were modeled on specimens collected from the native habitat of the birds, and the arrangement was done after drawings. Brunchorst noted that the glass cases of wrought iron and glass were very elegant. "Quite surprisingly beautiful" was the verdict. Again, the labels were emphasized as making this an outstanding "textbook" of British ornithology.[37]

The overall theme of Brunchorst's entry to the yearbook was that a collection should be like a textbook, with the labels providing the content, and the objects serving as illustrations. Tellingly, in many instances he would use the term "collection" when talking about a "display." Still, most of the material of the museum would be on display, even if he worked to put a new division into place, dividing the public from the research departments. He would also emphasize how carefully chosen samples from the collection should be made visually accessible to the public, while the scientist needed well-systematized collections of objects that were easily accessible for handling but also well protected against light. In this way the specimens would survive longer.

Two years later, Brunchorst presented, in a draft to his peers, a plan for the reorganization of the natural history department of Bergen Museum.[38] The new order he wanted was based on his ambition to make the objects more instructive for "the most numerous public to the museum," as he said, while making more space available for the collection. The first problem was that animal fossils were currently placed between other animals and animal remains. These would need to be displayed together to show the development of the animal forms. Further, he wanted to remove the foreign specimens that were placed within the collection of Norwegian fauna. He also emphasized the need to remove the doublets, which he found in abundance, from the bird collection. In cupboards on top of the existing cabinets there would be room for doublets and bird skins in light, tight conditions. Through such a reorganization, one could save one-third of the exhibition space used for the birds, according to Brunchorst, and one would in addition make the display more informative and perspicuous for the visiting public. For the specialist, there would be many opportunities to find the doublets in close proximity to the exhibited birds. This would even make room for a collection of Norwegian eggs that could be placed either together with the birds or in a desk covered by oilcloth to prevent them being damaged by the light. The main tenet of this plan was to divide the collection in two, with the scientific collections divided from the pedagogical exhibits. Some objects would be "textbook material," others the basis for research.

Brunchorst had listened carefully to leading international voices, not least of which was the director of the natural history department of the British Museum, William Henry Flower. Flower proclaimed in an 1889 address to the British Association for the Advancement of Science, "I believe that the main cause of what may be fairly termed the failure of the majority of museums—especially museums of natural history—to perform the functions that might be legitimately expected of them is that they nearly always confound together the two distinct objects which they may fulfill [research and instruction], and by attempting to combine both in the same exhibition practically accomplish neither."[39] For Flower, putting a complete collection on display was like framing and hanging on the walls all the book pages of the British Library; hence he called for a strict separation between public and scientific collections.

The most pertinent points of Flower's argument adhere to the status and being of the natural objects in these two different realms. The research collection should allow for careful investigations of the objects, and the objects should be treated as books in a library, as references.[40] In the public gallery, on the other hand, the number of objects should be limited, "according to the nature of the subject to be illustrated and the space available." Again, as with Brunchorst, we see how the natural object here acquires a different meaning. In the public collection, the exhibition object, the one that is to be seen in the glass case, has an altogether different value—as an illustration. The glass case is like an illustrated book where the text carries the intended meaning that the objects illustrate. "Above all," writes Flower, "the purpose for which each specimen is exhibited, and the main lesson to be derived from it, must be distinctly indicated by the labels affixed, both as headings of the various divisions of the series, and to the individual specimens. A well-arranged educational museum has been defined as a collection of instructive labels illustrated by well-selected specimens."[41] Here Flower cites the powerful and influential museum spokesperson and assistant secretary to the United States National Museum George Browne Goode, who was particularly keen on labeling objects, but also on stressing the pedagogic potential of object lessons.[42]

Given the division emphasized by Brunchorst, following up on Flower, the question is whether we can talk of two different kinds of museum nature, as two types of natural specimens distinguishable by their appearance in the public space of the museum as opposed to the research space. The answer is self-evidently yes: in the public space there should be specimens prepared in the best way, artistically and with a strong focus on their appearance. In the research department on the other hand, the objects come in two types: first, the valuable type specimens that have formed the basis for the description of the species, and second, as part of series in which the individual object is interesting in its minor differences from the next. The major difference between what objects are made to be in the two spaces lies principally in the distance created by the unavailability of the objects in the public part of museums. Because of the unflinching use of glass to cover the objects, they are also rendered permanently inaccessible to the visitor. In the display case, nature emerged as already identified and known. In the storage room, nature would make room for uncertainty. In the first instance, the objects are stilled and frozen, whereas a working collection would be dynamic as curators moved objects in and out of the cases to work with them, or entered new objects into the museum. In the public department, the object is made singular; in the research department, the objects are turned into part of a series. In the aforementioned texts, our two protagonists put the singular display object in view for the public and insisted on a pedagogy of standardization: to make an instructive and aesthetically pleasing experience for the visitor, the one single specimen would give the standard. The zoological specimen would serve as an exemplary illustration for a particular species. The type specimen in the research department would be the singular authentic object, while the series of specimen would make up researchable series—still in the research department. Maybe we can deduce that the singular object would stand for an idea of nature, whereas the working object would "be" nature.

## Reading Glass Cases

The glass cases were the means that could ensure such a separation of the collection, allowing the necessary visualization. But how exactly should a glass case be made, what were the principles on which it should be built, and what was the right material for building it? These museological questions were connected closely to new visions for museums, they were recorded in journals and in the museum yearbooks, and they were debated at international exhibitions and when museum staff visited one another's institutions.

Early in 1898, Bergen Museum's two new wings were almost finished, and the director was eagerly gathering information on how to furnish their new rooms. Brunchorst relied on letter writing, approaching German firms to obtain information about how much iron cases would cost. Iron cases were becoming something of a fad in the museum world, based on a design developed by the museum in Dresden. The museum director, Adolf Bernhard Meyer, publicized the design and the functionality of the cases in numerous publications.[43] The cases were in use at least in Vienna and Berlin in addition to Dresden, but the number of admirers spread much wider.[44] One of the principal drawbacks of these cases was the price; it was also difficult for local museums, using local carpenters, to produce such cases themselves. The American anthropologist George A. Dorsey was unimpressed by the iron cases, which were "lacking in beautiful finish," as well as "cumbersome" and "of enormous weight."[45] As far as we can ascertain, Bergen Museum did not purchase iron cases at this time, and we can only guess that this was because of the high price. In the absence of the fashionable iron models, one had to rely on wooden cases, and specifically homemade wooden cases. Information about how to make these cases could also be found in contemporary journals. Brunchorst addressed the Smithsonian thus: "In the report for 1893 of the U.S. National Museum, which is a real treasure to every museum-administrator, it is said, that photographs and working drawings of cases are lent to other museums. . . . It would be of very great interest to our museum . . . if you would kindly lend us some photos & working drawings."[46] Brunchorst had read the description of the museum cases developed at the United States National Museum, in their own report, and he hoped to receive usable drawings and photographs of them. Again we do not know whether he actually received the drawings, but reading the pamphlet and examining the photographs in the report of the United States National Museum certainly would have given him food for thought.

This report was written by Goode, who, together with Flower, came to epitomize the new museum politics and pedagogy in the last decades of the nineteenth century. He was able to paint a heaven of visionary museum ideas over basic museum work, to turn museum cases into active educators: "Each well-arranged case with its display of specimens and labels is a perpetual lecturer, and the thousands of such constantly on duty in every large museum have their effect upon a much larger number of minds than the individual efforts of the scientific staff, no matter how industrious with their pens or in the lecture

FIG. 2.2 The Whale Hall at Bergen Museum, c. 1895. Picture Collection, University of Bergen.

room." To fulfill this duty, the cases would have to be of glass, "the very best of glass in the largest possible sizes."[47] There should be as little wood as possible, and even the top—no matter what its size—should be of glass. Maximum glass would enable light to fall on the objects in the largest possible degree.

"The theory" that had led to the development of the cases Goode presented in the accompanying drawings was based on reading objects as texts: "Collections should be so arranged that each surface of glass, or each panel of a long case, stands by itself, its contents being grouped with reference to a general descriptive label, either placed in their midst or in the middle of the case-frame above." The manner of reading the case should be from left to right, and each panel should stand for itself, "like the page of a book." This approach to material culture is, of course, familiar to us today; we are used to theories that have read material artifacts as signs and texts. What is surprising is that Goode, who has become famous for his object-based approach to learning, leaned so heavily on a textual approach to museum display. But he relied, as did Flower and Brunchorst, on the idea that education, which was the museum's fundamental idea, needed words for transmission, and that in the public museum the objects functioned as mere illustrations of the knowledge that one could obtain from the written word. Thus visible objects and translucent cases were important for making the public read. Museum labels would be the companion to better museum cases. "The art of label writing is in its infancy," wrote Goode, "and there are doubtless possibilities of educational results through the agency of labels and specimens which are not as yet at all understood."[48] In Bergen Museum, we can follow staff making new labels in the various parts of the museum year by year; Goode's message had been received.

Considering leading nineteenth-century figures of American and British museums in the context of work undertaken in Bergen highlights the exchange of museum practices and the standardization of display that occurred in the period. The difficulty is in showing how international calls for universalized practices affected what the museums actually did, and how glass cases did their work in different localities. It is, for example, difficult to imagine that visitors came to read texts and look at the exotic animals as mere illustrations of the labels, as Goode had intended.

## The Case in Practice

What Brunchorst admired the most in London was the habitat group of British birds mentioned earlier in the essay. But in Bergen, there would be no room for similar displays of extensive habitat groups. In the years to come, and within the new wings, the collection nevertheless mushroomed. This was true of the historic-antiquarian collection, but even more so of the natural history collection, where more and more objects were cramped into existing glass cases and cabinets.[49] Examining pictures from the collections as well as reading the annual reports from the different departments tells the story of the amount of work that went into getting the display in order (fig. 2.3). Here one reads about constant refurbishing and installation of new cases, about extensive labeling programs and the movement of objects from display to storage areas. The program of making the display into a well-ordered textbook was on its way, but the sheer amount of specimens coming into the museum made the job difficult. The problem was also one of priority; there were intense discussions about the stated aim and future for the museum, and about where the emphasis should be placed—on developing the museum as an educational institution for "everyman," or on making the museum the basis for a new

FIG. 2.3 Interior from Bergen Museum, ca. 1925. From Professorkollegiet, *Bergen Museum, 1925*. Picture Collection, University of Bergen Library.

university in Norway (and, of course, whether one had to choose between the two). In the end, Brunchorst resigned his post in 1906, after intensive internal fighting, to become a diplomat in the newly independent Norwegian state.

Was the glass case, then, a means for enforcing the standardization of zoological specimens, or was it the other way around? Did the zoological specimens require specific cases? And what about all the zoological objects that did not go into cases, even as their display was also standardized? In fact, what is most striking in some of the early pictures from within Bergen Museum is the number of objects that were outside glass cases: skeletons of large animals and trophies, but also some large stuffed animals. Maybe they were too big, or maybe they were just so obviously "museum nature" that they did not need a glass case to position the visitor in the correct viewing perspective.

## The Enemy of Possession

In his essay "Experience and Poverty," Walter Benjamin writes of glass that it is "such a hard, smooth material to which nothing can be fixed."[50] It is a material on which there can be left no traces, according to him. Trying to focus my gaze on the glass cases in Bergen, Benjamin's description seems perfectly valid.

I have looked for traces of history, and for traces of the animals that were kept inside—the specimens that were left in the cases to be read as books. What I find are the letters and correspondence in the archive, the articles in the journals, and I read them. And then I look at pictures, but find them difficult to decipher. What work do the glass cases actually do? Andrew Zimmerman has written about glass cases in anthropological museums in the same period, pointing out that they were utilized for one special purpose: as a machinery to make anthropological objects into non-art, "as objects of natural science in the first place."[51] Looking at the pictures from Bergen, I can easily agree that the animals in the cases seem to be installed within a machinery that transforms them into objects of natural science, but this is a truism. They were regarded as natural science all along; in fact, they were part of the machinery that made glass cases into tools for rendering something as natural science.

What is more disturbing is the difficulty of seeing the animals within the glass cases in the various pictures. The plates of glass in the cases act as barriers that lock the animals up, making them not just untouchable for our hands but inaccessible for our gaze. Isobel Armstrong argues that glass was just as important as a barrier to sight as it was a medium for sight in the Victorian age. That is to say, there was a great interest in what glass did to our vision, and how it could prevent, as well as enable, good looking. Looking at the many pictures of museum galleries from this period, the degree to which the glass cases have become the most important features of the rooms is striking. They prevent our investigation of the objects, throwing themselves on us with an insistence of their particular materiality. The great paradox is that they were installed in the museums to make museum nature visible and legible for the greatest possible number of people. In the photographs, they refuse us access to this museum nature. "Glass is, in general, the enemy of secrets," Benjamin wrote, continuing, "It is also the enemy of possession."[52]

## NOTES

1. Scheerbart, *Glasarchitektur*, 56.
2. Alberti, "Constructing Nature behind Glass," 74.
3. Berger, "Why Look at Animals?" 5.
4. For the history of the museum, see Brunchorst, *Bergen Museums historie*; Professorkollegiet, *Bergen Museum*; Eriksen, *Museum*.
5. The building was finished for an international fishing exhibition in 1865, but the museum itself inaugurated the premises in 1867. See Brunchorst, *Bergen Museums historie*.

6. See in particular Flower, "Museum Organisation"; and Goode, "Recent Advances in Museum Method."

7. Armstrong, *Victorian Glassworlds*, esp. 89–92.

8. Heesen, "Vom Einräumen der Erkenntnis," 91.

9. Philipp Hainhofer of Augsburg organized the construction of a number of these for the nobility, and the most famous of them, the *Kunstschrank* of Gustav II Adolf, can be seen in Uppsala today.

10. Heesen, "Vom Einräumen der Erkenntnis," 92.

11. Ibid., 93–94.

12. Pearce, "William Bullock," 18–19.

13. Ibid., 15.

14. See Armstrong, *Victorian Glassworlds*.

15. Bennett, "Exhibitionary Complex."

16. Bonhams 1793, *Auction Catalogue, the Contents of Mr. Potter's Museum*.

17. Henning, "Anthropomorphic Taxidermy and the Death of Nature."

18. Nansen to Danielsen, February 22, 1886, in Bergen Museum Naturhistorisk avd. VIII D a 2. Brev 1886–1890, Statsarkivet i Bergen.

19. The first curator of the zoological department, Johan Koren, was hired in 1846; he was to be the only curator until 1876, when a second curator was hired. It was this post that Nansen earned in 1882. See Brunchorst, *Bergen Museums historie*.

20. Nansen to Danielsen, February 22, 1886, in Bergen Museum Naturhistorisk avd. VIII D a 2. Brev 1886–1890, Statsarkivet i Bergen.

21. This is obvious from the archives I have consulted, whether in Bergen Museum or at the American Museum of Natural History in New York.

22. Whale skeletons hunted from Norwegian stations are to be found in many European museums, some of them traded from Bergen Museum, some of them through other museums and dealers.

23. See the first months of 1888 in Bergen Museum. VIII C a 1 Brevjournal (for naturhistorisk avdeling) 1888–1893, Statsarkivet i Bergen.

24. See, for example, Grewe, *Die Schau des Fremden*.

25. For a contemporary source, see the preface to Hornaday, *Taxidermy and Zoological Collecting*, vii: "The rapid and alarming destruction of all forms of wild animal life which is now going on furiously throughout the entire world, renders it imperatively necessary for those who would build up great zoological collections to be up and doing before any more of the leading species are exterminated. . . . If the naturalist would gather representatives of all these forms for perpetual preservation, and future study, he must set about it at once."

26. Letter from Henry H. Giglioli, July 22, 1893, in Bergen Museum Naturhistorisk avd. VIII D a 3, Brev 1891–1893, Statsarkivet i Bergen.

27. For an elaboration on this theme, see Haraway, "Teddy Bear Patriarchy."

28. Alberti, *Nature and Culture*, 91.

29. Nansen to Danielsen, February 22, 1886, in Bergen Museum Naturhistorisk avd. VIII D a 2. Brev 1886–1890, Statsarkivet i Bergen.

30. Brandt to Danielsen, December 7, 1893, in Bergen Museum Naturhistorisk avd. VIII D a 3, Brev 1891–1893, Statsarkivet i Bergen.

31. Brunchorst, "Indberetning fra den naturhistoriske afdeling," 47.

32. Bergen Museum Naturhistorisk avd. Forhandlingsprotokoll 1852–1900. VIII A 1, Statsarkivet i Bergen.

33. Brunchorst, "Indberetning fra den naturhistoriske afdeling," 43.

34. The activity in the museum required more income, and the three chief sources of income for the museum were direct funding from the state, which made up half of the budget, with the rest divided between funds from Bergen Savings Bank, *Sparebanken*, and the Alcohol Cooperation. This cooperation, *Brennevinssamlaget*, was a fund raised by taxes on the production of alcohol from the 1870s. It was a local tax, made to use for local purposes.

35. Letter from Jørgen Brunchorst to the director of Bergen Savings Bank, February 23, 1894, in Bergen Museum Styret, II B b2 Kopibok 1894–1898, Statsarkivet i Bergen.

36. Brunchorst, "Om South Kensingtonmuseet," 15, 19, 20.

37. Ibid., 18.

38. Bergen Museum Naturhistorisk avd. VIII D a 3, Brev 1891–1893, Statsarkivet i Bergen.

39. Flower, "Museum Organisation," 15. The essay was based on a speech given in 1889.

40. The principal points to be aimed at in the research collection, according to Flower, were "the preservation of the objects from all influences deleterious to them, especially dust, light, and damp; their absolutely correct identification, and record of every circumstance that need be known of their history; their classification and storage in such a manner that each one can be found without difficulty or loss of time; and, both on account of expense as well as convenience of access, they should be made to occupy as small a space as is compatible with these requirements." Ibid., 16.

41. Ibid., 17, 18.

42. See *Annual Report of the Board of Regents of the Smithsonian Institution*.

43. See Meyer, "2. Bericht über einige Neue Einrichtungen."

44. See Dorsey, "Notes on the Anthropological Museums of Central Europe," 471, who didn't care for the iron cases himself, but emphasized that they were widely admired.

45. Ibid.

46. Bergen Museum Styret, II B b2 Kopibok 1894–1898, Statsarkivet i Bergen.

47. Goode, "Recent Advances in Museum Method," 23.

48. Ibid., 23, 37.

49. Professorkollegiet, *Bergen Museum*; *Norsk biografisk leksikon* (1999–2005), entry on Jørgen Brunchorst.

50. Benjamin, "Experience and Poverty," 733–34.

51. Zimmerman, "From Natural Science to Primitive Art," 287.

52. Benjamin, "Experience and Poverty," 734.

## BIBLIOGRAPHY

Alberti, Samuel J. M. M. "Constructing Nature Behind Glass." *Museum and Society* 6 (July 2008): 73–97.

———. *Nature and Culture: Objects, Disciplines, and the Manchester Museum*. Manchester: Manchester University Press, 2009.

*Annual Report of the Board of Regents of the Smithsonian Institution, Showing the Operations, Expenditures, and Condition of the Institution for the Year Ending June 30, 1897*. New York: Arno Press, 1980.

Armstrong, Isobel. *Victorian Glassworlds: Glass Culture and the Imagination*. Oxford: Oxford University Press, 2008.

Benjamin, Walter. "Experience and Poverty." In *Selected Writings: Vol. 2 (1927–1934)*, edited by Michael W. Jennings, Howard Eiland, and Gary Smith, 731–36. Cambridge, Mass.: Belknap Press, 1999.

Bennett, Tony. "The Exhibitionary Complex." *New Formations* 4 (Spring 1988): 73–102.

Berger, John. "Why Look at Animals?" In *About Looking*, 3–28. London: Writers and Readers, 1980.

Bonhams 1793. *Auction Catalogue, the Contents of Mr. Potter's Museum of Curiosities Tuesday 23 and Wednesday 24 September 2003, Jamaica Inn, Bolventor, Cornwall.*

Brunchorst, Jørgen. *Bergen Museums historie*. Bergen, 1900.

———. "Indberetning fra den naturhistoriske afdeling." In *Aarsberetning for 1894*, 43–52. Bergen, 1894.

———. "Om South Kensingtonmuseet." In *Aarbog for Bergen Museum*, 15–21. Bergen, 1891.

Dorsey, George A. "Notes on the Anthropological Museums of Central Europe." *American Anthropologist* 3, no. 1 (1899): 462–74.

Eriksen, Anne. *Museum: En kulturhistorie*. Oslo: Pax, 2009.

Flower, William Henry. "Museum Organisation." In *Essays on Museums*, edited by William Henry Flower, 1–29. London: Routledge, 1998 [1898].

Goode, George Brown. "Recent Advances in Museum Method." In *Annual Report of Board of Regents of the Smithsonian Institution . . . for the Year Ending June 1893*, 23–58. Washington, 1893.

Grewe, Cordula, ed. *Die Schau des Fremden: Ausstellungskonzepte zwischen Kunst, Kommerz und Wissenschaft*. Stuttgart: Franz Steiner Verlag, 2006.

Haraway, Donna J. "Teddy Bear Patriarchy: Taxidermy in the Garden of Eden, New York City, 1908–36." *Social Text* 11 (Winter 1984/85): 19–64.

Heesen, Anke te. "Vom Einräumen der Erkenntnis." In *Auflzu: Der Schrank in den Wissenschaften*, edited by Anke te Heesen and Anette Michels, 90–97. Berlin: Akademie Verlag, 2007.

Henning, Michelle. "Anthropomorphic Taxidermy and the Death of Nature: The Curious Art of Hermann Ploucquet, Walter Potter, and Charles Waterton." *Victorian Literature and Culture* 35 (2007): 663–78.

Hornaday, William T. *Taxidermy and Zoological Collecting: A Complete Handbook for the Amateur Taxidermist, Collector, Osteologist, Museum-Builder, Sportsman, and Traveller*. New York: Charles Scribner's Sons, 1894.

Meyer, A. B. "2. Bericht über einige Neue Einrichtungen des Königlichen Zoologischen und Anthropologisch-Ethnographischen Museums in Dresden." In *Abhandlungen und Berichte des Königl: Zoologischen und Anthropologisch-Ethnographischen Museums zu Dresden*, 4:1–25. Berlin: R. Friedlaender, 1892/93.

Pearce, Susan. "William Bullock: Inventing a Visual Language of Objects." In *Museum Revolutions: How Museums Change and Are Changed*, edited by Simon Knell, Suzanne MacLeod, and Sheila Watson, 15–27. London: Routledge, 2007.

Professorkollegiet. *Bergen Museum, 1925: En historisk fremstilling*. Bergen: Museets styre, 1925.

Scheerbart, Paul. *Glasarchitektur*. Berlin: Verlag der Sturm, 1914.

Zimmerman, Andrew. "From Natural Science to Primitive Art: German New Guinea in Emil Nolde." In *Die Schau des Fremden: Ausstellungskonzepte zwischen Kunst, Kommerz und Wissenschaft*, edited by Cordula Grewe, 279–300. Stuttgart: Franz Steiner Verlag, 2006.

**3**

## Preserving History: Collecting and Displaying in Carl Akeley's *In Brightest Africa*

*Nigel Rothfels*

. . . an everlasting monument to the Africa that was, the Africa that is now fast disappearing.
—CARL AKELEY

In his 1923 memoir, *In Brightest Africa*, Carl Akeley describes his collecting adventures in Africa working for the Field Museum in Chicago and the American Museum of Natural History in New York in the 1890s and first decades of the twentieth century. At first glance, the book seems to sit easily beside other hunting memoirs of the period, as Akeley relates his experiences shooting elephants, lions, antelopes, gorillas, and other animals in the "Dark Continent." Akeley did not consider himself a hunter, however, but a scientific collector, taxidermist, and artist. And, indeed, even though such figures as Theodore Roosevelt, Carl Georg Schillings, and many other big game hunters also characterized their hunting activities as quests to collect for natural historical museums, Akeley's memoir remains importantly different from their classic sport-hunting works. Despite its hunting focus, Akeley's account is clearly structured around the paradoxical theme of *preserving* animals at a time when their possible extinction seemed imminent.

The opening lines of the foreword to *In Brightest Africa*, by Henry Fairfield Osborn, reflect much of the tone of this endeavor. Osborn writes, "This is the daybook, the diary, the narrative, the incident, and the adventure of an

African sculptor and an African biographer, whose observations we hope may be preserved in imperishable form, so that when the animal life of Africa has vanished, future generations may realize in some degree the beauty and grandeur which the world has lost."[1] Osborn concludes that Akeley's commitment to truthfulness in his descriptions through his visual "biographies" of animals, "combined with his love of beauty of the animal form—beauty of hide, of muscle, of bone, of facial expression—will give permanence to [his] work, and the permanence will be the sure test of its greatness."[2]

According to Osborn, what makes Akeley's account important is its focus not on *killing* animals but on *saving* them. For most readers, though, this may seem a surprising claim, because the core of *In Brightest Africa* remains the telling of adventures shooting this or that species. Still, while most of the adventures in Akeley are not that different from those to be seen in the larger genre of hunting stories—after all, the basics of the hunt (reconnoitering the prey, stalking, searching out the best specimen, getting the right shot, locating the dead animal, and then finally taking stock of the animal and the hunt as the trophy is measured) are pretty much the same in all these works—it remains clear that Akeley's book is, in fact, quite different from the others. The distinction is important to make because, first, it puts the claims of sport hunters who were *also* museum collectors in needed perspective, and, second, it illuminates what can be called the "preservation paradox" that motivated people like Akeley—that killing, photographing, and pickling animals should be understood as practices of preserving them, of saving them.

Before turning to how Akeley's account is unusual, it is useful to see just how much his writing about hunting parallels the writings of the great hunters of the period, including such figures as Roosevelt, Schillings, C. H. Stigand, James Sutherland, Samuel White Baker, William Cornwallis Harris, Frederick Courtney Selous, Arthur H. Neumann, William Charles Baldwin, Hans Schomburgk, and many others. In what one might expect to be a particularly important story for his memoir, for example, Akeley relates the killing of the "old bull" at the center of the large elephant group in the American Museum in New York City. According to Akeley, he was hunting elephants one day in a forest that was high, thick, and dark, and realized that he was in an area where signs of elephants were everywhere. Looking up the trail, he thought he saw a group of the animals, but the shapes turned out to be just boulders. Moments later, though, he writes, "I saw across the gully another similar group of boulders, but as I peered at them I saw through a little opening in the leaves, plain and unmistakable, an elephant's tusk. I watched it carefully. It moved a little, and behind it I caught a glimpse of the other tusk. They were big and I decided that he would do for my group." Not able to see the animal's eye, Akeley calculated the point for a "brain shot" based on the location of the base of the tusk and fired. Akeley writes, "There was the riot of an elephant herd suddenly starting. A few seconds later there was a crash. 'He's down,' I thought." When Akeley reached the site with his gun bearer, though, the animal was gone, and they began to follow his trail, which "went straight ahead without deviation as if it had been laid by compass." Hours passed as the hunters followed the animal and Akeley notes that "the forest was so thick . . . we could not see in any direction." All of a sudden, Akeley writes, there was "a crash and a squeal," and the "elephant burst across our path within fifteen feet of us. It was absolutely without warning, and had the charge been straight on us we could hardly have escaped." As the animal quickly disappeared back into the forest, Akeley "fired two hurried shots." Realizing that the elephant had begun to stalk him and that he had just narrowly escaped from the wounded animal, Akeley "found a place a little more open than the rest" and decided to wait the animal out. He ate his lunch and had taken a couple of puffs on his pipe when the elephant "let out another squeal and charged." Akeley writes that he "didn't see him but [he] heard him, and grabbing the gun [he] stood ready." The elephant didn't come, though. Akeley concludes the story, writing, "Instead I heard the breaking of the bushes as he collapsed. His last effort had been too much for him."[3]

In many respects, Akeley's bull hunt echoes features of most classic elephant hunting stories of the period. He describes the difficult environment, the seeming impossibility of obtaining a clear shot, the imminent danger of a deadly charge, the dramatic moment when the hunter becomes the hunted, the almost overwhelming mental and physical challenge of trying to outwit and kill an elephant. These are standard elements of such stories, and they can be found throughout such abiding models of the genre as Samuel White Baker's *The Rifle and the Hound in Ceylon* of 1854 or Theodore Roosevelt's *African Game Trails* of 1910.

Baker's memoir of hunting during his time in Sri Lanka stands as one of the most deeply admired and best-known hunting accounts of the nineteenth century. And, like such works as Roualeyn Gordon Cumming's *Five Years of a Hunter's Life in the Far Interior of South Africa* from 1850, *The Rifle and the Hound* has been turned to repeatedly over the last 150 years as a rich source for both proponents and critics of sport hunting.[4] There are many tales of hunting elephants in the book, but they are told with an energy and graphic detail that became, in many ways, the hallmark of the genre. With a close attention to detail, these stories are intended to convey the excitement of hunting in the jungle for those who had never been there and to awaken thoughts and memories of similar great adventures for those who had. If they seem strikingly bloody and brutal today, the stories clearly shone with a very different luster for Baker and his avid readers. As he put it, for his own part, "These days will always be looked back to . . . with the greatest pleasure; the moments of sport lose none of their brightness by age, and when the limbs become enfeebled by time, the mind can still cling to scenes long past with the pleasure of youth."

A typical account in the book focuses on the hunt of a "cunning family" of three elephants, which had become targets for Baker because they were raiding crops. According to Baker, the elephants quickly scented the hunters and fled "down wind" through such a thick jungle that he was "very doubtful whether we should kill them." Nevertheless, Baker and his companions followed the animals through a mass of thorns, crawling at times on hands and knees. Baker writes, "I was leading the way, and could distinctly hear the rustling of the leaves as the elephants moved their ears. We were now within a few feet of them, but not an inch of their bodies could be seen, so effectually were they hidden by the thick jungle. Suddenly we heard the prolonged wh-r-r, wh-r-r-r-r, as one of the elephants winded us; the shrill trumpet sounded in another direction, and the crash through the jungle took place which nothing but an elephant can produce." The three animals split up, and Baker decided to follow the female, running after her for a half hour. She remained downwind, however, and always seemed to stay ahead of him. He writes:

> Speed was our only chance, and again we rushed forward in hot pursuit through the tangled briars, which yielded to our weight, although we were almost stripped of clothes. Another half hour passed, and we had heard no further signs of the game. We stopped to breathe, and we listened attentively for the slightest sound. A sudden crash in the jungle at great distance assured us that we were once more discovered. The chase seemed hopeless; the heat was most oppressive; and we had been running for the last hour at a killing pace through a most distressing country. Once more, however, we started off, determined to keep up the pursuit as long as daylight would permit.

Having run for two hours, Baker decided, because of the approach of dusk, that his only chance was to run faster, and he rushed after the elephant. Suddenly he found himself in a thick but low jungle, "through which no man could move except in the track on the retreating elephants." But then he saw the female at forty yards running quickly. In "the hopes of checking her pace," he fired at her ear. According to Baker, the elephant was now "inclined to fight, and she immediately slackened her speed so much that in a few instants [he] was at her tail, so close that [he] could have slapped her." Because of the thick jungle, though, Baker was stuck behind her and decided to fire his "remaining barrel under her tail, giving it an upward direction in the hope of disabling her spine."

Casting his empty gun aside, Baker reached and felt the "welcome barrel" of his spare gun pushed into his hand at the same moment that he "saw the infuriated head of the elephant with ears cocked charging through the smoke." When the smoke cleared, the elephant "lay dead at *six feet* from the spot where I stood. The ball was in the centre of her forehead, and B., who had fired over my shoulder so instantaneously with me, that I was not aware of it, had placed his ball within three inches of mine. Had she been missed I should have fired my last shot." With this, Baker pauses to contemplate the "glorious hunt": the great distance that he and his brother ("B.") had traveled, the remarkable fact that despite the distance they had only ended up three miles from their camp because the female had circled back, the disappointment that the bull and the younger elephant had "escaped," and the realization that shooting in thick jungles, especially because of the "obscurity occasioned by the smoke of the first barrel," is extraordinarily dangerous.[5]

Writing over fifty years after Baker and about adventures in Africa instead of Sri Lanka, Roosevelt's elephant-hunting adventures would nevertheless have been familiar to readers of Baker. When Roosevelt went to Africa, he was after Baker's elephants. Facing his foe, apparently alone much of the time, and preparing to shoot at the last possible moment, Roosevelt, too, describes a cunning and deadly beast. After a long stalk on a herd of elephants, "keeping ceaselessly ready for whatever might befall," for example, Roosevelt and his hunting companion Richard Cuninghame spotted "a big bull with good ivory." Aiming at a spot near the eye that he thought would lead to the brain, Roosevelt writes, "I struck exactly where I aimed, but the head of an elephant is enormous and the brain small, and the bullet missed it. However, the shock momentarily stunned the beast. He stumbled forward, half falling, and as he recovered I fired with the second barrel, again aiming for the brain. This time the bullet sped true, and as I lowered the rifle from my shoulder, I saw the great lord of the forest come crashing to the ground." The story doesn't end here, though. Roosevelt continues:

> At that very instant, before there was a moment's time in which to reload, the thick bushes parted immediately on my left front, and through them surged the vast bulk of a charging bull elephant, the matted mass of tough creepers snapping like packthread before his rush. He was so close that he could have touched me with his trunk. I leaped to one side and dodged behind a tree trunk, opening the rifle, throwing out the empty shells, and slipping in two cartridges. Meanwhile Cuninghame fired right and left, at the same time throwing himself into the bushes on the other side. Both his bullets went home, and the bull stopped short in his charge, wheeled, and immediately disappeared in the thick cover. We ran forward, but the forest had closed over his wake. We heard him trumpet shrilly, and then all sounds ceased.[6]

Roosevelt and Baker clearly belong to that school of hunters who somehow always manage the hairbreadth escape, firing desperate shots while miraculously throwing themselves to the side of the furious beast with bloodlust in its eye. These are adventure stories that are meant to quicken the pulse of the reader and show the hunter to be utterly fearless in the face of unimaginable terror. But while it is true that Baker and Roosevelt describe themselves and their hunts somewhat more enthusiastically than Akeley did his, the differences between these three hunts—with Akeley coolly puffing on his pipe while waiting for the elephant's charge, and Baker and Roosevelt firing at essentially point-blank range—do not seem that great. In a very real sense, they are all working within a constricted narrative space where the seemingly insignificant human goes up against the most powerful and deadly creatures, apparently one-on-one (despite the presence of gun bearers, trackers, and other hunters). Beyond the stories, the whole hunting activity was, moreover, essentially the same among all these men. This is most evidently the case between Akeley and Roosevelt, who were hunting at the same general time and place (Akeley even records visiting Roosevelt's camp), and, in fact, both used the same professional guide to lead their safaris. To be clear, I am not arguing that the specific experience or particulars of these hunts are the same, but that these authors/hunters chose to portray their hunts similarly.

But for all the parallels in these accounts, Akeley insists that he was never really a hunter at all but a collector. Indeed, despite the fact that his book comprises the usual stories of hunting this or that animal, he also goes out of his way to argue that typical sport hunting was little more than bluster with devastating repercussions on the animals of Africa. Indeed, with few qualms about scientific collecting, Akeley nevertheless seems to have had some general reservations about the "pleasures" of hunting. This is particularly clear in his account of collecting Somali wild asses for the Field Museum in 1896.

Arguing that when it came to certain kinds of shooting he often felt "a great deal like a murderer," Akeley describes his attempts to collect the North African wild asses as "one of the worst" hunts of his first trip to Africa.[7] According to his memoir, Akeley's party left camp at three o'clock in the morning along with some camels to bring the collected specimens back. Finally, at about eight in the morning, they spotted a lone ass. He writes, "We advanced slowly. As there was no cover, there was no possibility of a stalk, and the chance of a shot at reasonable range seemed remote, for we had found in our previous experience that the wild ass is extremely shy and when once alarmed travels rapidly and for long distances." At two hundred yards, the animal spooked, but then it came back, apparently curious. Akeley and his companion fired. He writes that the animal was "hard hit . . . but recovered and stood facing us." Approaching closer and worried about losing the animal, the pair fired again. According to Akeley, the animal "merely walked about a little, making no apparent effort to go away. We approached carefully. He showed no signs of fear, and although 'hard hit' stood stolidly until at last I put one hand on his withers and, tripping him, pushed him over."

After a harrowing day during which he and his companion almost died for lack of water and eventually robbed a caravan of milk at gunpoint, Akeley describes shooting a second ass before nightfall. He writes, "Just at dusk the shadowy forms of five asses dashed across our path fifty yards away and we heard a bullet strike as we took a snap at them. One began to lag behind as the others ran wildly away. The one soon stopped and we approached, keeping him covered in case he attempted to bolt. As we got near he turned and faced us with great, gentle eyes. Without the least sign of fear or anger he seemed to wonder why we had harmed him." Apparently, the animal had only suffered a small wound high on its neck and Akeley felt it should have been able to run away. He concludes, "We walked around him within six feet and I almost believe we could have put a halter on him." Without describing how

FIG. 3.1 Collecting Somali wild asses, May 1896. Photo © The Field Museum, #CSZ5990_LS (Carl Akeley).

he killed the ass, Akeley writes simply, "We reached camp about midnight and I announced that if any more wild asses were wanted, someone else would have to shoot them."[8]

Akeley took a photograph of one of the dead asses being carried by a camel. The photograph is not reproduced in his memoir but is part of the lantern slide collection of the Field Museum in Chicago (fig. 3.1). Akeley took a great many photographs of dead animals in Africa. In the same Akeley materials, for example, there is a whole series of photographs of different parts of a dead lesser kudu and there is also a photographic series documenting the skinning and skeletonizing of an elephant. Akeley also took photographs of trees and shrubs, termite mounds, and local peoples. Some of the photographs were clearly taken to document the physical appearance of the animal so that Akeley could reproduce the appearance of the skin against veins and muscles when he prepared his taxidermy. Some of the photographs were taken to document plants and landscapes for reference when preparing the large museum dioramas for which Akeley would become famous. And Akeley undoubtedly used some of the photographs when he gave lectures to museum benefactors and other interested audiences. The photograph of the ass on the camel, a black-and-white image that was later colored by hand, is clearly not about documenting a dead body or a scene. Possibly used in his lectures, it is a highly composed image that resonates with the regrets Akeley describes in his text.

The image is arranged as a triangularly shaped triptych presented against a backdrop of mountains rising on each side and framing the central drama. On the left are two men, heads bowed before the dead ass. On the right are two other men looking directly at the dead animal. The body positions of both pairs echo each other. On the left, the men's legs are together, their bodies are turned slightly away from the camera, and their right arms are held a little forward; on the right, the men's legs are one in front of the other, their bodies are open to the camera, and their left arms are bent at the elbow. One of the men on the left holds a pair of binoculars; the other carries a case likely designed for a camera—the camera, of course, which is being used for the photograph. On the right, both figures attend to the camel—one holds the animal's lead, the other rests his hand on the camel's withers. The focus of the image is the dead ass strapped to the back of the camel, covered with blankets and grasses to protect its skin from the taut ropes. This is not a simple snapshot, but an image composed to tell a story about the death of the ass in a desert setting. If Akeley simply wanted to take a picture of the dead ass on the camel, he would not have positioned the bodies so carefully, he would not have the men carrying the guns, he would not have the optical equipment so prominently in the shot, he would not have paid such close attention to the overall structure of the image.

In a sense, the image is the opposite of a classic trophy shot. Whereas the trophy shot, such as the famous photograph of Akeley with the leopard he killed with his bare hands (fig. 3.2), attempts to tell the *heroic story* of a hunt—and that it was pursued according to the rules of the game—by carefully positioning the hunter, the animal, and often the weapons against a significant setting, here the hunter is absent and the guns and the setting tell a simpler and tragically banal story of death. Whereas gun bearers and trackers typically stand outside the frame of the trophy shot—that photograph, after all, is intended to tell the story of one man or woman against the animal—here the hunter is absent and the gun bearers and trackers stand solemnly next to the dead animal, appearing to both mourn and care for the creature. This is

FIG. 3.2 Carl Akeley with a leopard, August 1896. Photo © The Field Museum, #CSZ5974_LS (Carl Akeley).

not the scene described by so many sport hunters of the time concerning the "bloodlust" of the "natives," but a quiet scene depicting somehow the separateness of the animals, landscape, and people of Africa from the American hunter. Nor is this the serene scene Akeley eventually created with his taxidermy of the collected asses at the Field Museum, a scene that tries to re-create the living look of the animals with their dead bodies.

With this photograph and with Akeley's text relating his hunt of the ass, we should return to the elephant hunt described earlier and ask how that story is different from the Roosevelt and Baker accounts. Most notably, of course, where all three men describe the physical and mental strain of the hunt, the cunningness of the elephant, and the realization that they are being hunted by the creature, and where all three face the inevitable charge of the elephant only to be saved at the last moment by the animal's death, Akeley doesn't have to fire his last shot. The animal—echoing the stories of the asses—simply dies from earlier wounds. The critical thing to remember about these accounts is that they are all essentially fish stories. Yes, in all cases an elephant was likely killed, but how the authors choose to tell the story of that death, how they chose to *display their collections* in public, is central to understanding the motivations of these hunters. There was nothing, of course, stopping Baker from changing, omitting, or adding details about his hunts in his stories—the details, including shooting an elephant up her anus, *are there for a reason*, just as other details are present or not in Akeley's accounts. In the end, the least important part of telling a good hunting story is close attention to the facts of the hunt. As Roosevelt put it in an earlier work, "No great enthusiasm in the reader can be roused by such a statement as 'this day walked twenty-three miles, shot one giraffe and two zebras; porter deserted with the load containing the spare boots.'"[9] A good hunting story, and the exhibition of that story in a memoir, museum display, dinner conversation, or trophy photograph, is not really about the facts surrounding an animal's death, but the significance of that death itself. The photograph of the ass on the back of the camel, then, functions precisely like a trophy shot because it plays a pivotal role in explaining the significance of the animal's death to audiences back home. Unlike the typical trophy shot, however, the story here is of the tragic death of an animal, and this is the critical point about Akeley.

In the end, what distinguishes Akeley's work—whether in museum halls or *In Brightest Africa*—was his awareness of the possibility of tragedy and injustice in the death of an animal and his effort to *preserve life* and not simply *celebrate the moment of death*. However much, then, his memoir echoes themes in the writings of Baker, Roosevelt, and others, the account remains fundamentally different. Of course, there is irony in Akeley's belief that the only way he could save animals for the future was to kill them, and one should never confuse Carl Akeley with a late twentieth- or early twenty-first-century popular conservationist like David Attenborough or Jane Goodall! Akeley appears to have been convinced that the days of the large game animals of the world were numbered, just as he was convinced that he had the ability to preserve them for the future in his artistic taxidermy. Akeley did not want to simply store the bodies of animals in liquid spirits (the equivalent in his time to collecting genetic material); he wanted to somehow capture a moment of the life-ness of the animal through his taxidermy.[10] Working almost entirely with just the *surface* of the animal, and placing that over a frame and a form that would somehow bring that surface back to life, Akeley sought to freeze a moment of an animal's life, to save that animal and even the species from their apparently inevitable absolute destruction.

On October 4, 1916, Akeley wrote a brief letter to Henry Ward, director of the Milwaukee Public Museum, about his method for preserving elephant skin for his taxidermized mounts. In short statements he writes:

> Have got my best results with following solution
> 7/8 Gambier 1/8 Quebracho—by weight.
> Start with a solution of 4° (barkometer) gradually raise to 10° in two weeks—then raise to 24° during the next two weeks raising 4° or 5° at a time—4 or 5 days apart
> 24° seems to be about right to finish in and takes 6 to 8 weeks all told.
> The salt that is in the skin is useful & I sometimes add a little more.
> In adding the tan extract—stir liquor thoroughly—test with barkometer & raise the desired no of degrees—that is, when the 4° liquor first week has had the skin a day or two the salt will raise it to 10° or more—and of course only 4° of this is tan extract.
> When the skin has been in 3 or 4 weeks take it out—stretch on the beams (like the one S helped with here)
> Repeat this operation a couple of times—two weeks apart before it is finaly [sic] tanned.
> During tanning skin must be agitated twice a day.
> I think S. has data on oiling and final working. My results here are perfectly satisfactory.
>
> Very truly yours<br>Carl E. Akeley[11]

Using two plant extracts to create a tanning solution (liquor), a barkometer (a specialized hydrometer used to measure the density of tanning liquids),

FIG. 3.3 Kindergarten children in front of *The Fighting Bulls*, September 1954. Photo © The Field Museum, #RF78904.

salt, a structure of beams for stretching and drying the skins, an immense amount of physical effort, and quite a lot of time, Akeley was able to achieve "perfectly satisfactory" results preserving skin. As Rachel Poliquin and others have shown, by the end of the nineteenth century, taxidermists had become increasingly aware of the relentlessness of decay in previously preserved skins. In response, museum scientists like Akeley sought to develop new techniques for preserving animal skins, because they felt that they might be among the last people to have the opportunity to somehow stop the disappearance of the world's animals.

There is another layer of irony in all this, though—one stemming from the nature of the skins themselves. In 2003, the Field Museum undertook an elephant conservation project. I use the word "conservation" because it embodies precisely the same slippage that the word "preservation" held for

Akeley and his colleagues. Where we, at first sight, are likely to think of preserving animals as saving them from extinction, Akeley thought that as well, but he also thought, as we have seen, of just plain pickling. Similarly, conservation here does not point to something like mitigating human-elephant conflict in Africa, but repairing the slowly cracking taxidermized skins of *The Fighting Bulls*, one of Akeley's great taxidermic works (fig. 3.3). Built from the skins of elephants that Akeley and his wife brought back from Africa in 1907, *The Fighting Bulls* depicts two elephants in battle, capturing a vision that Akeley had not observed in nature but which he clearly felt demonstrated the sheer power and majesty of adult male African elephants. The work does not show Akeley as the fastidious chronicler of nature, but rather Akeley as Osborn described him, as the biographer of animals, as the man who sought to tell a lasting story in an imperishable form.

In 2003, though, the fighting bulls standing in Field Hall were about to enter their second century as taxidermized specimens. The children who saw them in the early years of the exhibit when it was in the Rotunda of the old Field Museum—what was previously the Fine Arts Building of the World's Columbian Exposition and is now the Museum of Science and Industry—are long dead. Their own children are mostly dead, as well. Throughout the hundred years, though, the skins have continued to slowly decay while changing humidity and other environmental conditions have also damaged them. However much Akeley's techniques succeeded in freezing time, decay inevitably persists.[12] Of course, cracks almost inevitably appear in large works of taxidermy; when they do, conservators are summoned and the cracks are repaired; the illusion is restored, if only temporarily (fig. 3.4). Today, people still walk around the exhibit and admire the animals, still stare into the creatures' perplexingly animated glass eyes. It is true that the bulls are no longer the dramatic focus of the hall; they stand a bit to the side now and are clearly not understood as the stunners they once were. It is also true that despite Akeley's and Osborn's ominous warnings, the African elephant can still be found ranging relatively freely (if ultimately managed and constrained) in Africa. Akeley's monument, his biography of the lives of elephants, however, persists. It is true that it is composed, in the end, of just very thin and thinning pieces of skin. But the deeper story of those traces of long-dead elephants—a story to be found both in Akeley's *In Brightest Africa* and in his great works of taxidermy—remains a testament to an artistic vision that continues to affect viewers today.

FIG. 3.4 Conservation of *The Fighting Bulls*, September 2003. Photo © The Field
Museum, #Z94375_13D (Mark Widhalm).

## NOTES

1. Akeley, *In Brightest Africa*, ix.
2. Ibid., xii.
3. Ibid., 42–44.
4. For further discussion of Cumming and Baker, see Rothfels, "Killing Elephants."
5. Baker, *The Rifle and the Hound in Ceylon*, 77–84.
6. Roosevelt, *African Game Trails*, 298.
7. Akeley, *In Brightest Africa*, 114. For other aspects of shooting the Somali wild asses, see Rothfels, "Trophies and Taxidermy," 134–36.
8. Akeley, *In Brightest Africa*, 115–18.
9. Roosevelt, *Outdoor Pastimes of an American Hunter*, 328.
10. My argument here builds from the analysis of Akeley and his photographic work by Mark Alvey in "Cinema as Taxidermy."
11. Quoted with permission of the Milwaukee Public Museum Archives, Box 1.2. My thanks to Susan Otto of the Milwaukee Public Museum. Thanks also to Mark Alvey at the Field Museum for alerting me to this letter, for his help in deciphering a number of key words, and for his collegial support and friendship over many years.
12. For a related discussion of taxidermy and time, see Kaldal and Rothfels, "Reflections on the Vitrine."

## BIBLIOGRAPHY

Akeley, Carl. *In Brightest Africa*. New York: Doubleday, 1923.
Alvey, Mark. "The Cinema as Taxidermy: Carl Akeley and the Preservative Obsession." *Framework: The Journal of Cinema and Media* 48, no. 1 (2007): 23–45.
Baker, Samuel White. *The Rifle and the Hound in Ceylon*. London: Longman, 1854.
Cumming, Roualeyn Gordon. *Five Years of a Hunter's Life in the Far Interior of South Africa: With Notices on the Native Tribes, and Anecdotes of the Chase, of the Lion, Elephant, Hippopotamus, Giraffe, Rhinoceros, &c.* New York: Harper Brothers, 1850.
Kaldal, Ingvild, and Nigel Rothfels. "Reflections on the Vitrine." *Art and Research: A Journal of Ideas, Contexts, and Methods* 4, no. 1 (2011), http://www.artandresearch.org.uk/v4n1/kaldal.php (accessed July 30, 2012).
Poliquin, Rachel. *The Breathless Zoo: Taxidermy and the Cultures of Longing*. University Park: Pennsylvania State University Press, 2012.
Roosevelt, Theodore. *African Game Trails: An Account of the African Wanderings of an American Hunter-Naturalist*. New York: Syndicate, 1910.
———. *Outdoor Pastimes of an American Hunter*. New York: C. Scribner's Sons, 1905.
Rothfels, Nigel. "Killing Elephants: Pathos and Prestige in the Nineteenth Century." In *Victorian Animal Dreams: Representations of Animals in Victorian Literature and Culture*, edited by Deborah Denenholz Morse and Martin Danahay, 53–63. Burlington, Vt.: Ashgate, 2007.
———. "Trophies and Taxidermy." In *Gorgeous Beasts: Animal Bodies in Historical Perspective*, edited by Joan B. Landes, Paula Young Lee, and Paul Youngquist, 117–36. University Park: Pennsylvania State University Press, 2012.

# PART II

AUTHENTICATING

# PART II

AUTHENTICATING

**4**

# The Pleasure of Describing: Art and Science in August Johann Rösel von Rosenhof's *Monthly Insect Entertainment*

*Brian W. Ogilvie*

Now, instead of a worm we consider ugly, we have a butterfly; instead of a crawling creature, a flying one; instead of an insect that lives on and eats the willow-tree, another one that dwells near flowers. But we must consider it more closely.
—AUGUST JOHANN RÖSEL VON ROSENHOF, *Monthly Insect Entertainment*

When the Nuremberg miniature painter August Johann Rösel von Rosen-hof (1705–1759) issued the first *Insecten-Belustigung* (*Insect Entertainment*), a quarto-sized copperplate engraving with a quarto sheet (eight pages) of accompanying German text, in 1741, he had little idea whether the German public would take pleasure in the display in image and text of insects' metamorphosis and behavior.[1] Insects were scarcely a new subject for the artist's pencil and the engraver's burin; Joris Hoefnagel (1542–1601) and his son Jacob (1575–ca. 1630) had painted and engraved them in the late sixteenth century, while the first illustrated natural histories devoted to insects had been published at the beginning of the seventeenth. Maria Sibylla Merian (1647–1717) had made a name for herself illustrating caterpillars and their metamorphoses, while only seven years earlier the first volume of memoirs on the history of insects by the French academician René-Antoine Ferchault de Réaumur (1683–1757) had appeared from the presses of the Académie des Sciences in Paris.[2] The Lutheran pastor Friedrich Christian Lesser (1692–1754) had published an *Insecto-Theologia* (*Insect Theology*) in 1738. Rösel could count on a certain level of interest in insects from his contemporaries. Yet still he was taking a gamble.

He need not have worried: the publication sold well enough for him to decide to issue every month another two insect engravings, each accompanied by the creature's description. By 1746 Rösel had produced enough of these images and texts to issue a compilation, the first of what would become the four volumes of the *Monatlich herausgegebene Insecten-Belustigung* (*Monthly Insect Entertainment*). Two more followed during Rösel's lifetime, in 1749 and 1755, while a final compilation appeared posthumously in 1762, edited by Rösel's son-in-law, Christian Friedrich Carl Kleemann (1735–89). Kleemann, himself an artist, continued the series for another three decades.[3]

The success of the *Insect Entertainment* underscores the importance of insects in the culture of Enlightened Europe. Rösel's work both drew on and reinforced a view that human beings could learn much from insects while at the same time finding recreation in their display. In order to explore what insects meant to Rösel and his contemporaries—how and why they studied, described, and depicted them—we begin with a brief overview of artists' and naturalists' approaches to insects from the late sixteenth century to Rösel's day. We then turn to how Rösel studied, hunted, and collected insects, for his working methods help us understand how and why he chose to display them in sober engravings accompanied by detailed text. Like his contemporaries, Rösel took particular interest in showy, colorful insects such as caterpillars, butterflies, and beetles; unlike many of them, his interests ranged more widely.

After exploring how Rösel displayed his insects to his audience, we will examine the aestheticization of insects in Rösel's publications and the Protestant natural theology that underpinned them. Both were powerful motivations for his work, and both contributed to its enthusiastic reception. As Rösel continued his *Insect Entertainment*, he occasionally adopted the tools of contemporary naturalists: the microscope and the dissecting table, for example, which he used to explore crayfish (classified among the insects by many of his contemporaries) and freshwater polyps. Nonetheless, both the title of the *Insect Entertainment* and Rösel's continued self-identification as a "miniature painter" on its title page suggest that even in Enlightened Germany, it would be premature to draw a sharp distinction between scientific and artistic modes of knowing and displaying insects.

## Rösel and His Predecessors

Rösel did not begin his career intending to study and display nature. Scion of a noble Austrian family, the Rösels von Rosenhof, who had emigrated to

Franconia during the Reformation and then fallen on hard times, Rösel had been trained as an artist by his cousin and then worked as a miniature painter and engraver before obtaining a position as court painter in Copenhagen.[4] In the fall of 1727, on a journey to Amsterdam, he fell ill in Hamburg; during his convalescence, he happened upon a copy of Maria Sibylla Merian's folio volume *The Metamorphosis of the Insects of Surinam*.[5]

As Rösel later told the story, Merian's stunning, hand-colored engravings filled him with a desire to devote himself to studying and illustrating the insect world. Like other conversion narratives, Rösel's story of sudden enlightenment is open to question. Even his son-in-law and biographer, the artist Christian Friedrich Carl Kleemann, doubted that Rösel had undergone a sudden, miraculous transformation: the papers from his apprentice days included many studies of insects and animals. Rösel himself wrote that "from my youth I found enjoyment in insects, and paid close attention to the differences between caterpillars."[6] Merian's book may only have catalyzed a decision that was long in the making, but its effects were clear: Rösel gave up his voyage to Amsterdam and returned instead to Nuremberg, where he settled on his plan to produce the *Insect Entertainment*.

Merian herself was by no means the first European artist to be fascinated by insects. As Marcel Dicke has shown, insects appear in medieval Western art, but they become much more frequent in the seventeenth century.[7] In many of these cases, insects were ornamental details or appeared as part of a still life. But some artists from the late Renaissance onward took a more particular interest in insects.[8] The self-taught miniaturist and court artist Joris Hoefnagel filled the empty pages of handwriting sample books belonging to Emperor Rudolf II with exquisite miniatures of insects, amphibians, and flowers. His series of miniatures on "The Four Elements" included a set, "Ignis" (Fire), whose subject was "Reasoning Animals and Insects."[9] In 1592, Hoefnagel's son Jacob (and other engravers) produced the *Archetypa studiaque patris Georgii Hoefnagelii*, a series of engravings after Joris's model books, with emblematic or allusive quotations, in which insects were a central theme.[10]

These engravings and others modeled after them circulated widely and inspired further artistic imitation. In particular, one of the engravings in the *Archetypa* may have encouraged a significant development. Under the heading "Nasci. Pati. Mori" (birth, suffering, and death), the younger Hoefnagel arranged several dead or dying creatures along with a numbered sequence of caterpillar, pupa, and imago of the Spurge Hawk Moth (*Hyles euphorbiae* [L.]).[11] For millennia, careful observers had known that some caterpillars and "worms" transformed themselves into winged insects, but as far as I know, this is the first pictorial representation of insect metamorphosis. It may have been this engraving that inspired the first serious student of insect metamorphosis: an artist, Johannes Goedaert (1617–1668) of Middelburg in the Netherlands. For decades, Goedaert collected caterpillars, raised them in captivity, observed their metamorphoses, and described them in word and picture. He gathered these observations in a series of three small octavo volumes, published between 1660 and 1669, that combined his engravings with written descriptions.[12]

While Goedaert was observing and delineating his insects, the young Merian was doing the same. The daughter of the famous Nuremberg engraver Matthäus Merian the Elder (1593–1650), who died when she was only three, Merian was trained in drawing and painting by her stepfather, Jacob Marrel (ca. 1613–1681). From a young age, as she wrote in the preface to her *Metamorphosis of the Insects of Surinam* (1705), she was fascinated by silkworms and butterflies, and began to draw them and observe their changes. Her stepfather encouraged her to draw insects on plants as part of her artistic training, but the careful observation, she wrote, was at least partly due to a lack of friends or playmates.[13] Somehow she found a copy of Goedaert's book. Taking solace in the fact that her pursuit was not unique, and learning from Goedaert's technique of raising caterpillars in jars, Merian laid the groundwork for an impressive series of studies of insects and their metamorphoses: *Der Raupen wunderbare Verwandelung* (two volumes, 1679 and 1683) and the *Metamorphosis insectorum Surinamensium* (1705), each translated and produced in numerous editions.

These artists were participating in a wave of enthusiasm for insects that also included practitioners of natural history. Conrad Gessner (1516–1565), Thomas Penny (ca. 1530–1589), Jacob Zwinger (1569–1610), and Ulisse Aldrovandi (1522–1605), among other sixteenth-century naturalists, gathered material for the natural history of insects, though Aldrovandi was the first to publish any of his work. Aside from Aldrovandi's 1602 *De animalibus insectis* and the *Theatre of Insects* compiled by Thomas Moffett (1553–1604) and published posthumously in 1634, there were few works written by naturalists on insects before the middle of the seventeenth century. That would change dramatically as publications by Jan Swammerdam (1637–1680), Francesco Redi (1626–1697), Marcello Malpighi (1628–1694), Steven Blankaart (1650–1704), and others flooded from the presses.[14]

Artists and naturalists communicated and collaborated on their discoveries. Goedaert's work was annotated by his friend, the Middelburg physician

Johannes de Mey (1617–1678), who was actually mentioned as a co-author of the work by one seventeenth-century writer.[15] Merian discussed her work with Caspar Commelin (1668–1731), other Amsterdam physicians, and the famous collector Levinus Vincent (1658–1727). Naturalists, in turn, hired artists to illustrate their works, or to reproduce other illustrations, as with the Yorkshire physician Martin Lister's (1639–1712) reorganized translation of Goedaert.[16] Naturalists and artists exchanged information indirectly, too, avidly reading and critiquing one another's work, privately or in print. In Rösel's day, no student of insects was so specialized that he or she could ignore other experts, regardless of their background and training.

## Insect Student, Insect Hunter, and Insect Collector

Rösel, too, worked closely with scholarly experts on insects. He attended lectures in German by the Altdorf professor Michael Adelbulner (1702–79), and he discussed his work frequently with Georg Leonhard Huth (1705–1761), a Nuremberg physician and naturalist. Huth translated passages from Réaumur and other authors for Rösel, who could not read French. Rösel also read works by other German insect students: above all, Lesser's *Insecto-Theologia* and the series on German insects by Johann Leonhard Frisch (1666–1743).

Rösel's work benefited from his reading and his conversations with Huth. But it was primarily as a careful, patient observer that he impressed his contemporaries. Like Goedaert and Merian, Rösel was an adept insect hunter. As he tromped through the woods, fields, and meadows around Nuremberg, he kept his eyes open for unusual creatures, especially larvae: being wingless, they were easier to catch, and they would allow him to observe the insect's entire life cycle. He carefully noted the plants on which each insect lived.[17] After observing a creature's behavior in the wild, he would collect it, bring it home, observe it, and draw it in preparation for another installment of the *Entertainment*. Even when he had already carefully studied a particular insect species, though, he continued to observe it and note its behavior.

As more and more installments of the *Insect Entertainment* continued to appear, Rösel began to receive reports and specimens of peculiar insects from Nuremberg and farther abroad. But his reputation as an insect hunter could also hinder his access to unusual finds. In 1745 he received a drawing of a large caterpillar that was unknown to him and that he could not find in any insect book. He thought it might be the Death's Head Moth (*Acherontia atropos* [L.]) described by Réaumur, but there were significant differences between the drawing and Réaumur's illustration. His correspondent asked him to return the drawing, but before doing so Rösel made a copy.[18]

The following year he heard that a woman who had a garden near Nuremberg had found a beautiful caterpillar and was showing it for money: "My desire for new insects that I did not know swiftly led me to seek out this woman, but when she realized that I was the person to whom several of her supporters had asked her to bring the caterpillar, she did not want to show it to me. It cost me many flattering words to get a look from some distance. Even from afar, I recognized it immediately as the caterpillar I had been seeking for some time; I did not rest until I had acquired it, with cash and a little flattery." In this case, Rösel's reputation worked against him: fearing, rightly, that he would want to keep the caterpillar, the gardener avoided him and, when he found her, tried as long as possible to keep the insect from him. In vain: Rösel acquired it, and later that summer he was brought two further specimens.[19]

The drawback to receiving specimens from others, Rösel admitted, was that he could not necessarily learn every detail about the insect's life cycle. He confessed his ignorance of the early larval instars of the Death's Head Moth, because all three live caterpillars he received had already molted for the last time. Nonetheless, he was able to describe their pupation and metamorphosis into the splendid imago with eerie skull-shaped markings. In another case, he did not even have that experience. The Privet Hawk Moth (*Sphinx ligustri* [L.]), Rösel confessed, "is the first caterpillar I have described that I have not seen myself, though I have made every effort imaginable to find it."[20] Privet, he explained, is rare around Nuremberg. However, he decided to publish a description anyhow, based on material that he received from correspondents: a drawing of the caterpillar and pupa, sent to him from Lübeck along with an adult insect, and a description of the insect prepared by Paul Heinrich Gerhard Moehring (1710–1792), a physician based in Jever (Friesland). Rösel reproduced both caterpillar images, commenting on the differences between them as matters for further research. And he prepared engravings of two adults, sent to him through "the kindness of friends" to enlarge his collection.

Perhaps embarrassed by publishing a description of an insect he had not seen and studied while it was alive, Rösel lapsed into the subjunctive: "This would be the tenth moth in the second class," he wrote, "but I know still more, though I have not been able to find out whether their caterpillars are found around here: perhaps, though, I will find them, and then I will spare no effort to describe them and their metamorphosis."[21] Rösel published a secondhand description for several reasons: to please the patrons and friends who had sent him material, to entertain his readers—and perhaps also to make the month's deadline. But his goal was still to carefully describe behavior and metamorphosis that he had witnessed himself.

## Displaying Insects in the *Insect Entertainment*

Rösel conceived of his descriptions as a union of image and text; each referred to the other. And though some aspects of the *Insect Entertainment*'s balance between image and text derived from the technical constraints of printing, Rösel chose to deviate in significant ways from the work by Merian that had served as his inspiration. Both books were laid out so that images were displayed next to the text that commented on them, but Rösel's work did so while containing substantially more text for each image.

By the early eighteenth century, artists had developed a sophisticated theoretical discourse written by and for practitioners.[22] From the fifteenth century in Italy, humanists and artists like Leone Battista Alberti (1404–1472), Lorenzo Ghiberti (1378–1455), and Leonardo da Vinci (1452–1519) reflected on the nature of artistic imitation of nature. Intellectual and artistic contacts between Italy and transalpine Europe brought those currents north in the late fifteenth and sixteenth centuries. Artistic academies soon followed.[23] While apprenticeship continued to be a significant way for artists to learn their craft, the theory of the academies filtered into ordinary artistic activity. Rösel, for instance, apprenticed to his cousin but then completed his education in the Nuremberg Academy. By the end of his studies he was well aware of debates about the nature of art, the active imitation of nature by the artist, and the effects of art on its audience, and he must have considered those debates as he organized the *Insect Entertainment*.

The results of his reflection can be seen in the divergence between Rösel and his immediate inspiration, Merian's work on Suriname. Both artists intended their engravings to be studied along with the accompanying text, and they understood that their readers would move actively between word and image. But they made different choices about the proper balance between the two. Merian chose a format without too much text. As she put it in her letter to the reader, she kept her descriptions short so they could be set opposite the illustrations, following the example of Bidloo's anatomy. As a result, they occupy at most a folio page. She referred those who wanted more to books by Moffett, Goedaert, Swammerdam, Blankaart, and others.[24] Hence text and image could be placed opposite each other, with engraved images bound facing the folio text. The sixty images correspond to sixty pages of description. Merian did not give a detailed life history of the insects, mentioning instead some of the most striking aspects of their form, growth, habits, or uses by Surinamese colonists, slaves, or Indians.

Rösel, on the other hand, accompanied each engraving with a quarto sheet of text. These sheets were originally published separately with the accompanying illustration, and Rösel's printer, Johann Joseph Fleischmann, fit the text to the sheets, either filling up blank space with ornamental flourishes or switching to smaller type in the last few pages of a sheet in order to fit in the entire text. Unlike Goedaert, Rösel (or Fleischmann) gave precise instructions to bookbinders on what to do with the engravings. His "Nachricht an den Buchbinder" instructed him not to simply bind the engravings along with the text, but, depending on what the owner wants, to either bind them together at the end of the volume or in among the descriptions. In either case, the engravings were to be tipped into a blank leaf so that they could be folded out and examined while reading the text.[25]

This format allowed Rösel to describe species at length while having the image constantly before the reader's eyes. And Rösel, unlike Merian, included in his engravings numbers or letters that were keyed to the text. Modeled after the keys that were common in works of anatomy and natural history, these cross-references linked the images closely to the accompanying descriptions.[26] Read and examined together, image and text formed the complete "insect entertainment," which might in turn inspire readers to seek out and observe the insects themselves.

## Caterpillars and Butterflies

As Rösel organized his work with its effect on his readers in mind, he also chose to begin with insects that were beautiful and fascinating. Butterflies and moths (Lepidoptera, in Linnaean terms) occupied a significant place in his and other early modern books on insects. Even in works that addressed many kinds of insects, like Hoefnagel's *Ignis* and *Archetypa* and Goedaert's *Metamorphosis*, butterflies and moths appear frequently. Rösel's immediate inspiration, Merian, placed them front and center. Her *Metamorphosis insectorum Surinamensium* is devoted primarily to butterflies: forty-eight of the sixty plates have a butterfly or moth as the central subject, while another two plates give equal emphasis to a lepidopteran and another subject.[27] In the end, Rösel would address a broad range of insects in his *Insect Entertainment*. But at first he had to build a subscriber base for his work; that may be why he began with the crowd-pleasing lepidopterans.

As Anita Albus has observed, butterflies pose a challenge to the artist: their vivid, often iridescent colors took skill and care to reproduce.[28] Their beauty only heightened the challenge. Moreover, to artists fascinated with insect metamorphosis, lepidopterans have the added attraction that their caterpillars were themselves often strikingly colored. It is not surprising, then, that they received so much attention. Finally, lepidopterans experience full (holometabolous) metamorphosis, undergoing a radical change as they pass from larva to pupa to imago. They were considered not only things of beauty in themselves but also instructive and even edifying, in ways that were inflected differently from the late Renaissance to the Enlightenment.

Hoefnagel's engraving "Nasci. Pati. Mori," for instance, appears to allude to the parallel between the caterpillar's transformation into a butterfly and the human being's metamorphosis from earthly to spiritual life.[29] This interpretation, and the notion of radical metamorphosis that underwrote it, was vehemently attacked by Jan Swammerdam, whose work on insects, conducted from the 1650s until his death in 1680, aimed to show the "insensible changes" that lay behind the appearance of radical transformation.[30] Rösel did not take an explicit position on this question. As a visual artist and observer, his investigations and depictions began with the external surfaces and behavior of the insects. But Rösel implicitly treated egg, larva, pupa, and imago as different forms of the same creature.

In depicting this life cycle, Rösel deviated from his predecessors in a few key areas. A striking feature of Merian's butterfly and moth illustrations was their ecological context: she portrayed each stage in the insect's life cycle on the plant that she thought served them as nourishment.[31] Rösel occasionally provided ecological settings, but in general he depicted the insect against a blank background, as did Hoefnagel and Goedaert. Unlike all three predecessors, however, his depiction of the insect life cycle was more detailed. Hoefnagel, Goedaert, and Merian (not to mention Ulisse Aldrovandi and other Renaissance naturalists) had divided the insect life cycle into three forms: larva, pupa, and imago.[32] Rösel, on the other hand, often provided multiple illustrations of distinct larval instars and, in some cases, enlarged anatomical details (fig. 4.1). Like Merian, he has also portrayed both the top and the underside of the butterfly's wings—with exceeding care, as can be seen by comparing one of his engravings with a modern photograph of the species.[33] He had their engravings hand-colored in his workshop, before sale, to ensure that the engraving matched, as closely as possible, the watercolor originals on which they were based.

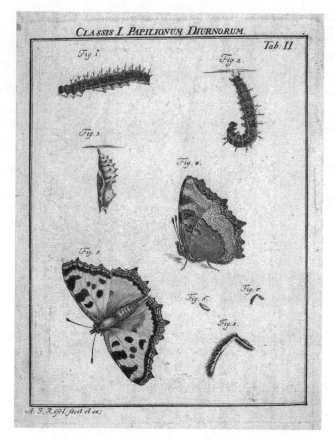

FIG. 4.1 Large Tortoiseshell butterfly. From Rösel, *Insecten-Belustigung*, volume 1. Heidelberg University Library, O 1314 RES::1, Teil1_Tab2.

## An Ordinate Fondness for Beetles

If butterflies and moths took pride of place in Rösel's *Insect Entertainment*, he also devoted substantial attention to beetles. This, too, is not surprising. An anecdote is often repeated about the British biologist (and atheist) J. B. S. Haldane. After a public lecture, the story goes, Haldane was approached by an audience member who asked him what his decades of studying nature had taught him about God. Haldane's response: "He has an inordinate fondness for beetles."[34] Haldane's remark was not random: beetles (coleoptera) constitute about 40 percent of insect species that have been described to date, and about 25 percent of *all* animal species.[35]

Like butterflies, many adult beetles are large, brightly colored, and iridescent. Beetles also had a significant place in northern Renaissance art, in part because of the admiration for Albrecht Dürer's (1471–1528) insect miniatures. His *Stag Beetle*, as Janice Neri has shown, played a central role in the Dürer Renaissance of the late sixteenth century, a movement in which Hoefnagel was deeply involved.[36] And like butterflies, beetles were the subject of a developed symbolism from antiquity and the Middle Ages: the Egyptian scarab, or dung beetle, is especially notable.[37]

Beetles range in size from tiny, nearly microscopic weevils to enormous tropical beetles. In the preface to his description of his first class of beetles, Rösel indicated this range with an engraving showing several different tropical rhinoceros beetles next to two small European dung beetles.[38] Not surprisingly, the largest and most elegantly colored species attracted artists' attention. The first engraving in Hoefnagel's *Archetypa* gave pride of place to the elephant beetle (*Megasoma elephas*) under the biblical quotation "Say to God, How terrible are thy deeds! So great is thy power that thy enemies cringe before thee," and above the inscription "To him who gave me the skill I shall give the glory."[39] Other beetles, including the stag beetle and many longicorn beetles, turn up in subsequent pages, though usually only the adult forms.

Goedaert and Merian showed relatively little interest in beetles. In Martin Lister's edition of Goedaert, only 15 of the 144 different "discoveries" involved beetles.[40] And though Merian's hand-colored engravings, based on original watercolors, were well suited to depict the splendor of certain beetles, she did so only infrequently.[41] Her attention, like Goedaert's, was drawn much more by splendid caterpillars than by drab beetle grubs.

Rösel, on the other hand, was fascinated by beetles. A substantial part of the second volume of the *Insecten-Belustigung* was taken up by beetles, divided into three classes of land beetles and one of water beetles.[42] As we will see, Rösel knew that some of his readers had little interest in insects other than butterflies, but he insisted that they were worth his, and their, attention. In some cases that came from their beauty or from the sheer magnitude of exotic species. But he was also fascinated by their life cycle, and in his descriptions he attempted to convey to his readers that fascination.

The first beetle Rösel described was, in fact, the cockchafer or Maikäfer (fig. 4.2). These well-known insects, he wrote, please everyone because after a harsh winter they are the sign of the advent of spring: "Therefore I hope that my efforts to give a complete report on the generation, growth, and metamorphosis of this insect will not be displeasing to my worthy readers. For, though there are few people in this part of the world who since their childhood have

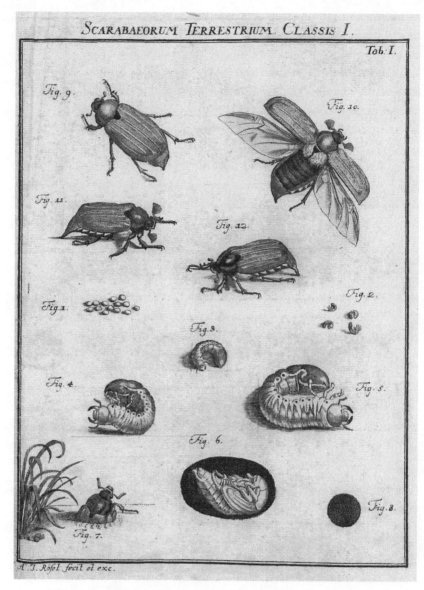

FIG. 4.2 Cockchafer. From Rösel, *Insecten-Belustigung*, volume 2. Heidelberg University Library, O 1314 RES::2, B_024a.

not been familiar with the cockchafer, there are very few who know how they are generated, how they grow, and how they transform themselves."[43] In this case, the "entertainment" comes from learning an unknown truth about a well-known creature.

As Rösel informs his readers, the cockchafer grows out of a grub (worm) that spends four years of its life underground. After mating, the female cockchafer lays a number of small eggs. Rösel collected some, without knowing what insect had laid them, placed them with soil in black jars, and put them in his cellar. The following spring they had hatched into small grubs that farmers knew as serious pests. He fed them first with grass and then with plant roots. As time passed they grew larger, and he moved them into clay pots. In the wild he had noticed several different sizes of such grubs, so to speed his identification of the adult insect he collected a few of the largest ones and put them in pots to observe. At the same time he kept the original grubs so he could figure out how long it took them to become adults.

The answer to that question, he reports, was four years. In the autumn of the third year the grub, now large, pupates, well below the surface of the earth (several feet deep in some instances). In the winter the adult hatches and slowly digs its way to the surface, appearing in the spring. Rösel noted that there were two commonly occurring forms of adult cockchafers in Franconia: some with a black thorax and some with a red one. They alternated years, so he concluded that each year's population was formed by the eggs that had been laid four years previously. He drew the further conclusion that, since the winter of 1740 had been long and hard, and few of the adult insects were able to emerge and mate, there would also be few adults in 1744.[44]

In this instance, as in many other passages in Rösel's work, we see the artist as naturalist. Rösel gathered up eggs, hatched them, and kept pots of them around, changing their food sometimes every other day, for four years, in order to determine with certitude what became of them. To study the metamorphosis more quickly he did collect grubs that were already in a late instar, but he kept the originals both to determine how long they remained larvae and to ensure that he had identified the proper transformation. Though he continued to identify himself on the title pages of successive parts of the *Insect Entertainment* as a "miniature painter," his comportment resembled that of many naturalists, both his contemporaries and his successors.

## The Aestheticization of Nature

What, then, distinguished Rösel from his contemporaries who might have identified themselves as naturalists?[45] Or rather, why did he choose to distinguish himself? Surely there was an element of advertising involved: the fine, hand-colored illustrations were the main selling point of the *Insect*

*Entertainment.* But Rösel responded vehemently to critics who suggested that he was not the actual author of his observations, and that he had merely provided illustrations to accompany a ghostwriter's words. Still, though, his approach to his subjects was shaped by his experience, as an artist, with aesthetic questions. His images, like those of other insect artists, were intended to evoke an aesthetic response.

In Western culture, one is tempted to say, nature is always already aestheticized through humans' immediate aesthetic response to it.[46] We may speak more narrowly, however, of aestheticization in the sense of taking *naturalia* and applying the canons of art to them. This, in turn, should be distinguished from the attempt to evoke an aesthetic response to nature in the reader. In the narrow sense, Rösel did not set out to aestheticize nature. Unlike Merian, he did not situate his insect larvae, pupae, and imagines in a carefully composed ornamental framework, though some of the illustrations in his 1758 *Natural History of Our Frogs* have a painterly composition.[47] Rather, his insect illustrations, carefully hand-colored by himself or under his supervision (as Kleemann notes, he would not have had time to do them all himself), are composed with an eye to filling out the engraving systematically, and using space well. In this case it is instructive to compare the engraved frontispieces to Rösel's first three volumes with the actual *Insect Entertainment* itself. The latter contains sober drawings, with the insects generally organized in blank space (corresponding, perhaps, to the artist's work table or his collection of dead insects). The frontispieces, on the other hand, are carefully composed. Volume 1's frontispiece is an allegorical composition by Johann Justin Preißler, engraved by Martin Tyroff (1704–1758). Rösel himself produced the frontispiece to volume 2 (fig. 4.3), while volume 3 features the work of Nicolaus Gabler (1725–1780), engraved by Michael Rößler (1705–1777). All three reveal that the sober style of the *Entertainment*'s engravings was deliberate.[48]

Despite this sober, technical style, Rösel clearly wanted to evoke an aesthetic response in his readers. He depicted beautiful creatures in a skillful manner. In this sense his work harks back to a 1597 cabinet miniature by Joris Hoefnagel that depicts insects and flowers with the motto "We take delight twice when we see the painted flower competing with the real: in one we admire the skill [*artificium*] of nature, in the other, the painter's ability [*ingenium*]."[49] But Rösel also depicted ugly creatures, and he was at pains to underscore that his work was not intended simply to please the senses: "Men have sharply differing inclinations: however, I hold that the more noble inclination should always be preserved. To love something only because it delights the

FIG. 4.3 Engraved title page to Rösel, *Insecten-Belustigung*, volume 2. Heidelberg University Library, O 1314 RES::2, I.

senses, without also directing one's attention to its Author or to the use that one can receive through it—that has never been what brought me to investigate insects. My intention has always been nobler; I have undertaken these investigations to praise the Creator and to be of use to my neighbor."[50]

Understanding and fully appreciating the pictorial display in his *Insect Entertainment*, Rösel implied, required knowledge of its order and purpose.

Naive enthusiasm for superficial beauty was a low form of pleasure; true understanding enhanced the overall aesthetic impact of displaying nature.[51] Rösel realized that his publication, with its textual excurses, would not please those "who wish only to see the bright butterflies in my collection," but he had received enough support and praise that he was not concerned about such critics.[52] As the reference to the "Author" of nature implies, the pious Protestant also hoped that his work, in evoking an aesthetic response, would at the same time provoke a religious response.

## The Insects' God

Like many other early modern students of insects, Rösel thought that his work should lead his readers to a deeper devotion to God. Insects taught one to admire God's divine workmanship in his creation. Rösel saw the sign of divine providence in the fact that butterflies that fed on nectar in meadows nonetheless deposited their eggs on the very different plants that the hatchling caterpillars would eat.[53]

The access to divine craftsmanship in the works of Rösel and his contemporaries was multiplex.[54] In a passage that could refer to Rösel, Heidrun Ludwig writes of Merian's *Raupenbuch* that it "opens three possibilities for devotion: (1) in considering and observing the visible world, (2) in active, artistic reflection of what is seen, and (3) in edification on the basis of the artistic reflection, whether in picture or word."[55] In other words, the artist him- or herself performs an act of devotion, both in observing the world and in accurately describing and depicting it. The reader, meanwhile, comes to know God better by contemplating the artist's work—which may, in turn, lead him or her to the study of insects themselves in nature. The artist's skill does not attempt to emulate God, as was the case with Joris Hoefnagel; rather, it is in the *fidelity* of artist to nature, and therefore to God, that the devotional practice lies.

Rösel certainly took his natural theology seriously. Though he never learned Latin or attended university, he studied William Derham's (1657–1735) *Physico-Theology* by attending Adelbulner's German-language lectures on the book.[56] And as Sara Stebbins has shown, Rösel's natural theology forms a minor current in a broad stream of German Protestant thought in the late seventeenth and eighteenth centuries. This "physico-theology" was characterized by increasing specialization on specific parts of the world, as opposed to the general harmony of things that was emphasized in classical and Renaissance natural theology. And it moved from carefully constructed arguments, in an apologetic mode, to evoking emotions of piety in its readers.[57] One of the leaders of this movement was Lesser, who wrote a *Litho-Theologia* (on stones) and a *Testaceo-Theologia* (on shells) in addition to the *Insecto-Theologia*. Lesser spoke highly of Rösel's *Insect Entertainment* in "Thoughts about the Insect-Entertainment," prefaced to the work's first volume: "You present to us, dear Rösel, many little creatures / Which in turn the Creator's hand made so wonderfully."[58]

### Rösel's Reception and Later Work

As Lesser's poem suggests, Rösel's combination of beautiful engravings, engaging descriptions, and conventional piety successfully reached a wide audience. The Danzig naturalist Johann Philipp Breyne (1680–1764) thanked Christoph Jacob Trew (1695–1769) for sending him some of the *Insect Entertainment*: they awoke in him a "particular pleasure," and they by far surpassed all the other illustrated works on insects. Johann Christian Müller, pastor of Reinsdorf near Zwickau, also underscored the "pleasure" that he "discovered in your *Insect Entertainment*." That pleasure was twofold: Müller read the monthly publication, but it also inspired him to follow Rösel's advice to investigate and experience insects and their transformations himself. "It is a special pleasure for me," he continued, "to show many people your beautifully colored engravings and, at the same time, the original in nature."[59]

Such praise, more than he had initially hoped for, encouraged Rösel to continue his enterprise. Some critics, it is true, thought that his descriptions were too long-winded, but Rösel defended himself by pointing out that he could sell the engravings alone for the price he asked for both illustration and text. Customers who were more interested in pretty pictures than in insect behavior had no grounds for complaint—they could simply ignore the text and be none the poorer.[60]

And as he continued his work, he expanded its scope. To the butterflies, moths, and beetles with which he began, he added other creatures: water bugs, dragonflies and damselflies, grasshoppers and crickets, bees and wasps, gnats and flies.[61] In the third, supplementary, volume, Rösel added even more insects: "the cunning, skillful Ant-Robber" (i.e., the ant lion), water spiders, gall wasps, and—in a reminder that the term "insect" had a broader meaning for Rösel than it does for modern zoology—two kinds of crayfishes, as well as a "History of Polyps and Other Water Insects."[62]

As Rösel took on new insects, he also expanded the techniques he used to study and portray them. By the late 1740s he was dissecting insects and portraying their internal structure as well as their external form. His study of the crayfish was his most sustained effort in "insect" anatomy, aided no doubt by the creature's size. But he also dissected beetles and, in homage to Malpighi, the silkworm larva.[63] The stroke he suffered in the early 1750s may have encouraged his interest in dissection, since it limited his field expeditions. In the beginning, Rösel followed the example set by artists from Hoefnagel to Merian, who had depicted the insect's exterior, leaving its interior structure to anatomists. By the end of his life, Rösel could wield a scalpel as well as he handled a burin.

As he continued, Rösel partially abandoned his decision, following Merian, to portray insects in their actual size. Most insect species that have been described have adult bodies that are shorter than five millimeters.[64] From the beginning, Rösel used magnifying glasses to examine insects' bodily structures; he even had a sun microscope that could project images on the wall of a darkened room, so he could show small insects to an audience. But only rarely in the first decade did he show enlarged details of legs, antennae, or scales. And he did not enlarge the entire insect. In this respect, his work differed from that of microanatomists, whose books devoted quarto- or folio-sized engravings to minute organs that were invisible to the naked eye. Only in his history of freshwater polyps did Rösel systematically enlarge his subjects. The title page to this section of his work showed the creatures life-sized, while subsequent illustrations, prepared with the aid of a hand lens or microscope, depicted them at different levels of magnification.[65] When he returned to other insects, he abandoned this systematic use of magnification. And he continued to work on butterflies and moths to the end, praising their beauty to his subscribers and readers. The fascination and wonder with insects of all sorts—even spiders— that characterized his earliest monthly entertainments never left him.[66]

## Conclusion

Rösel's *Insect Entertainment* is only one manifestation of the fascination with insects that characterized European culture from the late Renaissance through the Enlightenment. Merian's work, and that of other artists and naturalists, provided one spur to his studies, but insects themselves played a central role. Rösel was intrigued by the successive transformations of holometabolous insects, especially the larger, more colorful ones like butterflies, moths, and large beetles, but also grasshoppers, crickets, true bugs, and even spiders and slugs. He devoted time and patience to them. Rösel's jars and pots full of cockchafers took four years to cycle through, and though he speeded the work by finding older grubs, he still followed the original set through to its final transformation. He pursued his studies despite failing health. He was no dilettante.

From one perspective, Rösel's *Insect Entertainment* can be read, along with the works of his artist and naturalist predecessors and contemporaries, as part of the history of science: in particular, the history of natural history and entomology before their professionalization. In that framework, the contrast between Merian's lush, carefully composed engravings and Rösel's sober illustrations appears to mark a shift toward the modern scientific illustration. Rösel certainly has been adopted as an illustrious predecessor by modern entomologists: a cited reference search in the Web of Science database turns up nineteen articles that cite Rösel, all of them in scientific journals (published between 1907 and 2001).[67] And as I have argued elsewhere, naturalists, physicians, and artists were eagerly exchanging knowledge of nature even as they critiqued one another vehemently, thereby weaving artists' insect books into the complex web of early modern science.[68]

We should be wary, though, of imposing the modern notion of "science" on these investigators of nature, or of reducing the early modern fascination with insects to a stage in the production of systematized knowledge. As we have seen, Rösel was a careful observer, and contemporaries placed him among the "nature experts" (*Naturkündiger*) of his day. But such study was also a means to honor God by showing his providence and craftsmanship. And, it is worth noting, Rösel consistently identified himself on the title page of his works as "Miniatur-Mahler." He suffered many reproaches for his work, but the only one that struck to the quick, according to his son-in-law and biographer Kleemann, is the charge that he had not studied and could not read Latin or foreign-language works. This led him to attend courses and to seek the assistance of scholars like Huth for his work. But he insisted on the value of his work and, against his calumniators, maintained that it was his own: that *Insect Entertainment* was the result of his own experience, the experience of a miniature painter. By displaying insects, he demonstrated both his own talent in his chosen field and the beauty and moral lessons that the Creator had placed in nature's own miniatures, the insects.

## NOTES

1. Rösel, *Insecten-Belustigung*, unpaginated Vorrede, sig. [B4]r.
2. See the discussion below for details.
3. Kleemann, *Beyträge*. Rösel also published a natural history of frogs and toads in 1758: *Historia ranarum*.

4. Unless noted, biographical data are from the life by Rösel's son-in-law, C. F. C. Klee-mann, "Ausführliche und zuverläßige Nachricht." Some writers incorrectly assert that Rösel was ennobled in the 1750s because "von Rosenhof" first appears on the title page of part 3 of the *Insecten-Belustigung*. In fact, as Kleemann testifies, Rösel had merely secured an impe-rial confirmation of his ancestors' diploma of nobility (21–22).

5. Merian, *Metamorphosis*; facsimile of engravings in Merian and Schmidt-Loske, *Insects of Surinam*. See also Reitsma, *Merian*.

6. Rösel, *Insecten-Belustigung*, 3:86. Note: the pagination of Rösel's collected volumes of the *Insecten-Belustigung* is tricky. Part 1 contains six "collections," each paginated sepa-rately. Unless otherwise noted, all translations are my own.

7. Dicke, "Insects in Western Art."

8. The following brief synopsis is based on Ogilvie, "Nature's Bible." See also Vignau-Wilberg, "*In minimis maxima conspicua*"; Bertoloni Meli, "Representation of Insects"; and Neri, *The Insect*.

9. A selection of images from "The Four Elements," now in the collection of the National Gallery of Art in Washington, D.C., may be viewed online at http://www.nga .gov/cgi-bin/tsearch?oldartistid=202360&imageset=1 (accessed May 29, 2010).

10. Hoefnagel, *Archetypa*; a facsimile edition with commentary can be found in Vignau-Wilberg, *Archetypa*. An annotated copy from the Bibliothèque nationale et univer-sitaire de Strasbourg is available online at http://imgbase-scd-ulp.u-strasbg.fr/thumbnails .php?album=865 (accessed May 29, 2010).

11. Hoefnagel, *Archetypa*, part 2, no. 8.

12. According to unpublished research by Kees Beart, summarized by Ella Reitsma, Goedaert's three volumes were published in 1660, 1664/65, and 1669, but most publications continue to give 1662 and 1667 as the dates of the first two volumes. Reitsma, *Merian*, 68n22.

13. Merian, *Metamorphosis*.

14. See Bodenheimer, *Materialien zur Geschichte der Entomologie*; and Ogilvie, "Nature's Bible." I will not address these naturalists in this essay.

15. Reitsma, *Merian*, 68. The author was Christoph Arnold, who contributed a lauda-tory poem to Merian's first caterpillar book, published in 1679.

16. Goedaert, *Of Insects*.

17. For example, the moth he found on willow and linden trees. See *Insecten-Belustigung*, 3:67.

18. The story of the "unusually large Jasmine caterpillar, beautifully adorned with gold and blue, and its transformation into the so-called Death-Moth," appears in ibid., 3:5–16.

19. Ibid., 3:6, 7.

20. Ibid., 3:8, 25–26.

21. Ibid., 3:32.

22. For an overview, see Williams, *Art Theory*, 54–91.

23. On some interesting consequences of the introduction of academies, see Barkan, *Unearthing the Past*.

24. Merian, *Metamorphosis*, ad lectorem.

25. Rösel, *Insecten-Belustigung*, vol. 1, final unpaginated leaf. The copy in the Heidel-berg University Library, now available digitally, has the illustrations tipped into the left side of a blank leaf at the beginning of each description. The copy in the Yale University Library has the illustrations tipped into the right side of a blank leaf after each description.

26. On these keys, including their extensive use by Andreas Vesalius, see Kusukawa, *Picturing the Book of Nature.*

27. Merian, *Metamorphosis.* Plates 1, 18, 24, 27, 28, 48, 49, 50, 56, and 59 emphasize another insect, though butterflies or moths are present on some of them, while plates 21 and 54 seem to give equal emphasis to a lepidopteran and another insect. Plate 49 features the Lantern Fly, a true bug with colorful wings.

28. Albus, *Art of Arts,* 100–102, 290–91.

29. Hoefnagel, *Archetypa,* part 2, no. 8.

30. Swammerdam, *Bybel der natuure.*

31. Davis, *Women on the Margins,* 151–55.

32. See Ogilvie, "Nature's Bible."

33. See, for example, *Nymphalis polychloros,* the Large Tortoiseshell, Wikimedia Commons, photograph by user Algirdas, http://commons.wikimedia.org/wiki/File:Nymphalis _polychloros.jpg (accessed May 29, 2010).

34. The anecdote appears frequently, but without sources, in dozens of pieces on Haldane. Garson O'Toole (a pseudonym) traces its origin in print to Hutchinson, "Homage to Santa Rosalia." See O'Toole, "The Creator."

35. Bartlett, "Order Coleoptera."

36. Neri, *The Insect,* 5–10; Ogilvie, "Nature's Bible."

37. Cambefort, "Sacred Insect."

38. Rösel, *Insecten-Belustigung,* vol. 2, "Scarabaeorum Terrestrium Praefat: Classis I.," tab. A.

39. Hoefnagel, *Archetypa,* part 1, no. 1; translations by Vignau-Wilberg in *Archetypa,* 59–60.

40. Goedaert, *Of Insects,* 101–18.

41. In four out of sixty plates of the *Metamorphosis.* See Merian, *Metamorphosis,* 24, 28, 48, and 50.

42. Rösel, *Insecten-Belustigung,* vol. 2, four sections with independent pagination. Technically, the first class of water insects included those that metamorphosed into beetles; the second and subsequent classes were not beetles. Rösel's classification predates that of Linnaeus.

43. Ibid., "Der Erd-Kefer erste Classe," 2:1.

44. Ibid., 2:1–8.

45. On late Renaissance equivalents for the term, see Ogilvie, *Science of Describing,* 54–55.

46. For an overview on the aesthetics of nature in Western philosophy, see Parsons, *Aesthetics.*

47. On the context of Merian's work, including her flower books as well as her studies of insects, see Neri, *The Insect,* chap. 4.

48. Volume 4, published posthumously, has a portrait of Rösel as its frontispiece. Some of the engravings in the *Entertainment* depict water insects in a highly schematic ecological setting, normally showing the relationship between the aquatic and the terrestrial phases of their lives.

49. The miniature is now in the Muzeul Brukenthal, Sibiù, Romania; see Vignau-Wilberg, *Archetypa,* p. 32, fig. 9.

50. Rösel, *Insecten-Belustigung,* vol. 1, Vorrede, §4.

51. In this regard, Rösel appears to approach the position of the contemporary aesthetician Allen Carlson, but the difference is profound: whereas Carlson rejects the theistic

view that nature has a designer, emphasizing instead natural order, Rösel saw order as a consequence of divine design, as I will discuss below. However, both would agree that Noël Carroll's "being moved by nature" is an inferior form of aesthetic response. See Carlson, "Appreciating Art"; and Carroll, "On Being Moved."

52. Rösel, *Insecten-Belustigung*, vol. 1, Vorrede, §4.

53. Ibid., "Erste Classe der Tag-Vögel," 1:8.

54. On natural theology from the late Renaissance through the Enlightenment, see Stebbins, *Maxima in minimi*; Ogilvie, "Natural History"; Jorink, *Boeck der Natuere*; and Trepp, *Von der Glückseligkeit*.

55. Ludwig, *Nürnberger naturgeschichtliche Malerei*, 81–82.

56. Kleemann, "Ausführliche und zuverläßige Nachricht," 14.

57. Stebbins, *Maxima in minimis*, 12. Given that Stebbins situates these works within the framework of the "early Enlightenment," this move is ironic—at least if Enlightenment means rational criticism!

58. Rösel, *Insecten-Belustigung*, vol. 1, sig. E3r.

59. Ibid., vol. 1, sig. D4v–E1r.

60. Ibid., vol. 1, sig. B4r–v.

61. Ibid., vol. 2, contains these groups.

62. Ibid., vol. 3.

63. Ibid., 3:305–50; vol. 2, "Der Erd-Kefer erste Klasse," 57–72; 3:37–62.

64. Raffles, *Insectopedia*, 9.

65. Rösel, *Insecten-Belustigung*, 3:442 and fig. 72. Figures 73, 74, and 75 show one species of polyp life-sized and at two different levels of magnification. Rösel continued this procedure throughout the history of polyps. On early attempts to describe polyps and the frustration of analogy, see Elkins, "Visual Desperation."

66. The last six plates in the posthumous fourth volume are devoted to spiders. His son-in-law Kleemann wrote that they were the last plates and descriptions that Rösel prepared shortly before his death: Rösel, *Insecten-Belustigung*, 4:264.

67. ISI Web of Science database, search criteria: Cited Author=(roesel a* OR rosel a*) AND Cited Year=(1740–1780) Timespan=All Years. Databases=SCI-EXPANDED, SSCI, A&HCI, CPCI-S, CPCI-SSH (accessed May 31, 2010).

68. Ogilvie, "Nature's Bible."

## BIBLIOGRAPHY

Albus, Anita. *The Art of Arts: Rediscovering Painting*. Translated by Michael Robertson. New York: Alfred A. Knopf, 2000.

Barkan, Leonard. *Unearthing the Past: Archaeology and Aesthetics in the Making of Renaissance Culture*. New Haven: Yale University Press, 1999.

Bartlett, Troy. "Order Coleoptera." *BugGuide*, April 26, 2005. http://bugguide.net/node/view/60 (accessed June 3, 2010).

Bertoloni Meli, Domenico. "The Representation of Insects in the Seventeenth Century: A Comparative Approach." *Annals of Science* 67, no. 3 (2010): 405–29.

Bodenheimer, F. S. *Materialien zur Geschichte der Entomologie bis Linné*. 2 vols. Berlin: W. Junk, 1928–29.

Cambefort, Yves. "A Sacred Insect on the Margins: Emblematic Beetles in the Renaissance." In *Insect Poetics*, edited by Eric C. Brown, 200–222. Minneapolis: University of Minnesota Press, 2006.

Carlson, Allen. "Appreciating Art and Appreciating Nature." In *Landscape, Natural Beauty, and the Arts*, edited by Salim Kemal and Ivan Gaskell, 199–227. Cambridge: Cambridge University Press, 1993.

Carroll, Noël. "On Being Moved by Nature: Between Religion and Natural History." In *Landscape, Natural Beauty, and the Arts*, edited by Salim Kemal and Ivan Gaskell, 244–66. Cambridge: Cambridge University Press, 1993.

Davis, Natalie Zemon. *Women on the Margins: Three Seventeenth-Century Lives*. Cambridge: Harvard University Press, 1995.

Dicke, Marcel. "Insects in Western Art." *American Entomologist* 46, no. 4 (2000): 228–37.

Elkins, James. "On Visual Desperation and the Bodies of Protozoa." *Representations* 40 (Autumn 1992): 33–56.

Goedaert, Johannes. *Of Insects: Done into English and Methodized with the Addition of Notes*. Edited and translated by Martin Lister. York: Printed by John White for M. L., 1682.

Hoefnagel, Jacob. *Archetypa studiaque patris*. Frankfurt am Main, 1592.

Hutchinson, G. E. "Homage to Santa Rosalia; or, Why Are There So Many Kinds of Animals?" *American Naturalist* 93, no. 870 (1959): 145–59.

Jorink, Eric. *Het Boeck der Natuere: Nederlandse geleerden en de wonderen van Gods schepping, 1575–1715*. Leiden: Primavera Pers, 2006. Revised translation: *Reading the Book of Nature in the Dutch Golden Age, 1575–1715*. Translated by Peter Mason. Leiden: Brill, 2010.

Kleemann, Christian Friedrich Carl. "Ausführliche und zuverläßige Nachricht von dem Leben, Schriften und Werken des verstorbenen Miniaturemahlers, und scharfsichtigen Naturforschers, August Johann Rösels von Rosenhof." In Rösel von Rosenhof, *Der monatlich herausgegebene Insecten-Belustigung*, vol. 4. Nuremberg: Printed by Johann Joseph Fleischmann, 1762.

———. *Beyträge zur Natur- oder Insecten-Geschichte*. Nuremberg, 1761–92.

Kusukawa, Sachiko. *Picturing the Book of Nature: Image, Text, and Argument in Sixteenth-Century Human Anatomy and Medical Botany*. Chicago: University of Chicago Press, 2012.

Ludwig, Heidrun. *Nürnberger naturgeschichtliche Malerei im 17. und 18. Jahrhundert*. Acta Biohistorica: Schriften aus dem Museum und Forschungsarchiv für die Geschichte der Biologie, 2. Marburg an der Lahn: Basilisken-Presse, 1998.

Merian, Maria Sibylla. *Metamorphosis insectorum surinamensium, in qua erucae ac vermes surinamenses cum omnibus suis transformationibus ad vivum delineantur et describuntur, singulis eorum in plantas, flores et fructus collocatis in quibus reperta sunt, tum etiam generatio ranarum, buforum rariorum, lacertarum, serpentum, araneorum et formicarum exhibetur*. Amsterdam: Printed by Y. Valk, 1705.

Merian, Maria Sibylla, and Katharina Schmidt-Loske. *Insects of Surinam*. Hong Kong: Taschen, 2009.

Neri, Janice. *The Insect and the Image*. Minneapolis: University of Minnesota Press, 2011.

Ogilvie, Brian W. "Natural History, Ethics, and Physico-theology." In *Historia: Empiricism and Erudition in Early Modern Europe*, edited by Gianna Pomata and Nancy Siraisi, 75–103. Cambridge: MIT Press, 2005.

————. "Nature's Bible: Insects in Seventeenth-Century European Art and Science." *Tidsskrift for kulturforskning* 7, no. 3 (2008): 5–21.

————. *The Science of Describing: Natural History in Renaissance Europe.* Chicago: University of Chicago Press, 2006.

O'Toole, Garson (pseud.). "The Creator Has an Inordinate Fondness for Beetles." *Quote Investigator*, June 23, 2010. http://quoteinvestigator.com/2010/06/23/beetles/ (accessed November 28, 2010).

Parsons, Glenn. *Aesthetics and Nature.* Bloomsbury Aesthetics. London: Continuum, 2008.

Raffles, Hugh. *Insectopedia.* New York: Pantheon, 2010.

Reitsma, Ella. *Maria Sibylla Merian and Daughters: Women of Art and Science.* Los Angeles: J. Paul Getty Museum, 2008.

Rösel von Rosenhof, August Johann. *Historia naturalis ranarum nostratium: Die natürliche Historie der Frösche hiesiges Landes.* Nuremberg: Printed by Johann Joseph Fleischmann, 1758.

————. *Der monatlich herausgegebenen Insecten-Belustigung erster[-vierter] Theil.* 4 vols. Nuremberg: Printed by Johann Joseph Fleischmann, 1746–61.

Stebbins, Sara. *Maxima in minimis: Zum Empirie- und Autoritätsverständnis in der physikotheologischen Literatur der Frühaufklärung.* Frankfurt am Main: Peter D. Lang, 1980.

Swammerdam, Jan. *Bybel der natuure . . . of Historie der insecten tot zeekere zoorten gebracht.* Edited by Hermann Boerhaave. 2 vols. Leyden: I. Severinus, 1737–38.

Trepp, Anne-Charlott. *Von der Glückseligkeit alles zu wissen: Die Erforschung der Natur als religiöse Praxis in der Frühen Neuzeit.* Frankfurt am Main: Campus, 2009.

Vignau-Wilberg, Thea. *Archetypa studiaque patris Georgii Hoefnagelii 1592: Natur, Dichtung, und Wissenschaft in der Kunst um 1600: Nature, Poetry, and Science in Art Around 1600.* Munich: Staatliche Graphische Sammlung, 1994.

————. "*In minimis maxima conspicua*: Insektendarstellungen um 1600 und die Anfänge der Entomologie." In *Early Modern Zoology: The Construction of Animals in Science, Literature, and the Visual Arts*, 2 vols., edited by Karl A. E. Enenkel and Paul J. Smith, 1:217–43. Boston: Brill, 2007.

Williams, Robert. *Art Theory: An Historical Introduction.* Malden, Mass.: Blackwell, 2004.

**5**

Images, Ideas, and Ideals: Thinking with and about Ross's Gull

*Henry A. McGhie*

In culture, not all species are considered equal: some species and groups of species are more highly admired and desired than others, just as others are feared and detested to varying degrees. Some species are regarded differently by different groups of people. Some have rich cultural histories while others are more or less unknown, and there are "celebrity species" just as there are "celebrity animals" (such as the Saint Bernard dog Barry and Knut the polar bear, discussed in chapters 6 and 9, respectively). How do species develop a particular "reputation" or "persona" for groups of people? How do texts, visual representations, and museum specimens contribute to the development of that reputation? These questions form the substance of this essay and are explored through the cultural history of a single species of bird, Ross's Gull.

Ross's Gull is a small species of seabird that breeds in marshes in the northern tundra and winters at sea. Since first being made known to Western science in 1823, it has enjoyed an almost legendary reputation among ornithologists and specimen collectors. When the Norwegian scientist and Arctic explorer Fridtjof Nansen (1861–1930) first encountered the birds on his famous *Fram* expedition, he wrote the following in his diary: "Friday 3 Aug 1894. Latitude 81°5'. . . . Today at last my longing has been satisfied, I have shot Ross's Gull (Rodosthetia [*sic*] Rossii), and 3 of them in one day. This elusive, strange and

I am grateful to Eleanor MacLean (Department of Rare Books and Special Collections, McGill University) and to Mary Markey (Smithsonian Institution, Washington, D.C.) for providing me with copies of archival material and for granting me permission to quote from this material.

rarely seen inhabitant of the mysterious north, a world to which the imagination alone aspires and of which no one knows its coming and going, is that thing, from the first moment I saw these tracts and my eyes surveyed the lonely plains of ice, I had always hoped to discover."[1]

Since the time of its discovery, new information has been added about the bird's breeding biology, migration, and vagrancy, although much remains only poorly known. It is known to breed in northeast Siberia, locally in Greenland, and irregularly in Canada. Non-breeding adult gulls are known to occur in the Arctic Ocean all the way to the North Pole in summer. Its winter quarters remain unknown but probably lie near the edge of pack ice.[2]

The species has long been of interest to naturalists and collectors, and continues to elicit a high level of interest among many modern-day birdwatchers. Preserved specimens of the bird were extremely rare in collections and were exhibited and re-exhibited at scientific meetings in Britain, notwithstanding that there was little to say about them. During the nineteenth-century "culture of collecting," specimens of the rare bird and its eggs were not available to collectors. The species appeared frequently in collectors' lists of "desiderata"; when specimens did eventually become available (from the end of the nineteenth century for skins and the early twentieth century for eggs) they were traded among leading European and American naturalists for high values, discussed further below. James Fisher, who popularized birdwatching in Britain in the mid-twentieth century, could still write of it as "one of the most mysterious birds in the world."[3] Similarly, Michael Densley, who wrote a monograph on the species, considered that "the bird's association with the Arctic and its early explorer-heroes, its rarity and elusiveness and, not least, the beauty of its plumage, have endowed it with a fascination for ornithologists probably unequalled by any other species."[4] Russian and American ornithologists have expressed similar views.[5] More recently, similar views have been expressed by birdwatchers blogging about their quests to see rare birds. A straying individual bird that visited Massachusetts in the 1970s was described by one blogger as "the bird of the century."[6]

This essay uses Ross's Gull as an animal to think about and to think with: as real animal, standardized animal, idea, and aesthetic. This is, in essence, a study of influences, of the power of words, pictures, and word-pictures, and of people and human relationships, on the development of ideas about the natural world. By examining the development of the "reputation" of this particular species through a variety of texts, images, and locations of production (notably in museums and elitist societies), this investigation seeks to establish how "they" became an "it" through a combination of scientific and cultural practices. These practices are investigated in detail to explore how, more generally, ideas and knowledge about animals develop. Ross's Gull is especially well suited to this type of investigation, as its distinguished reputation has resulted in a large body of published detail being available about its scientific description, naming, preservation, dissemination, and popular culture, detail which is not available for most animal species. Particular emphasis is placed on the development of ideas about the bird in the nineteenth century.

The development of ideas about animals is linked with their scientific discovery. A caricature of scientific discovery could run something like this: Someone acquires the remains of a dead (usually) animal or plant. A naturalist encounters these remains and considers them different enough from similar species to constitute a new species not yet described in published literature. Someone publishes an article giving the new species a two-part scientific name (genus and species) accompanied by a description of the new species' distinctive characteristics by which it (or some members of it) can be recognized. The specimen(s) forming the basis of the description, called type specimens, are usually deposited in a museum collection where they are preserved for reference. Over time, new information about the species becomes known and is published. Museums and private collectors may acquire more specimens. The species may develop a popular history and feature in travel narratives and field guides. People may choose to search for the species themselves, for study or pleasure. Through this essay, I will explore how this caricature is a gross oversimplification of the complexity of human-animal relationships; how many of the episodes are far more complex and contingent; and how these contingencies influence the development of ideas about animals.

## Making a Name for Yourself

Ross's Gull was first made known to Western science in 1823 when two specimens were shot in Greenland. This event took place during the course of William Parry's (1790–1855) second voyage in search of the Northwest Passage, part of Britain's imperial program of exploration. The two birds were shot by James Clark Ross (1800–1862) and one of his companions. One of these two birds (the bird shot by Ross) was described scientifically the following year. The preserved skin of this bird is now to be found in a drawer in the National Museums of Scotland in Edinburgh; the remains of the other bird are in World Museum in Liverpool. The specimen in the National Museums of Scotland is of more than ordinary interest to scientists as, apart from its historical attachments, it is also a type specimen, the named and described basis of a whole species of bird, the standard by which all other similar (and sometimes not so similar) individual birds are assessed and classified. Type specimens are "the foundation of a taxonomical pyramid that links the individual . . . to the . . . kingdom through ascending taxa of ever greater generality; each level of the hierarchy is typified by a designated representative at the next lower level: the species by the type specimen, the genus by the type species, and so on."[7] The Edinburgh specimen is the holotype ("main" or "only" type) of the name *Larus roseus*. Ross's Gull currently goes by the name *Rhodostethia rosea*, scientists having decided it is different enough from other gulls to warrant its own genus; the name "rhodostethia" refers to the pinkish breast.

The scientific naming of Ross's Gull is associated with controversy. In October 1823 the collections from Parry's second Arctic and earlier voyages were handed over to specialists to catalog and describe; the two "new" gulls were included among these collections. Birds were to be described by the naturalist John Richardson (1787–1865), a Scottish surgeon-naturalist and veteran Arctic explorer. He recognized the gull as belonging to a new species and planned to call it *Larus rossii* in order to honor its discoverer, basing his description on one of the specimens but mentioning the other, making them a holotype ("main" type) and paratype ("supplementary" type).[8] It was standard practice at the time to name species in honor of someone associated with them: the collector, owner, purchaser, or some other patron. In so doing, Richardson also connected his own name with that of the beautiful new species and with that of Ross, who was a national hero.[9] James Clark Ross lived for almost forty years after his discovery of the species. He was the archetypal Polar hero, the first to reach the North Magnetic Pole (he also tried for the South); through his lifetime "Ross's Gull" existed as a kind of personal possession of this famous celebrity and society figure. From the time of its first description, Ross's Gull has had a particularly strong association with its discoverer, whose exploits feature in many accounts of the bird and its history.

Richardson announced his new species at a meeting of the Wernerian Society in Edinburgh. He held back from publishing his description of the new gull so that the description could form part of the official account of the expedition, a more illustrious place of publication. Another leading Edinburgh-based naturalist, William MacGillivray (1796–1852), gave the new bird a temporary scientific name (apparently based on only one of the specimens), making clear that it was to be properly described elsewhere.[10] However, the temporary name was used by some other naturalists and was adopted as the bird's proper name, resulting in bad feeling from Richardson toward Mac-Gillivray. After this time, the so-called law of priority—that the first published name associated with a description had to be adopted for a species—became standard practice, meaning that MacGillivray's temporary name had precedence over Richardson's intended, but unpublished, scientific name.

The naming affair and the bad feeling it caused toward MacGillivray has been described several times.[11] However, claims that MacGillivray acted unfairly, as he was aware that his newly created scientific name would be the one that would have to be used in perpetuity, are unfounded. Scientific nomenclature was more fluid at that time, and the law of priority had not yet been fully developed or accepted.[12]

Through the nineteenth century, the scientific name of the bird was standardized as *Rhodostethia rosea*. The name *Rossia* (again in commemoration of James Clark Ross) was proposed as a new genus but this had already been given to a type of squid, and so couldn't be given to another genus of animal. Several authors continued to ignore scientific conventions and tradition, and used what to them was the more appropriate name of *Larus rossii* or *Rhodostethia rossii*, to commemorate the bird's links with its illustrious discoverer.[13] Vernacular names are not standardized in the same way and are more reflective of cultural trends. Through the nineteenth century the bird was known as the Cuneate-tailed Gull, Wedge-tailed Gull, or the Rosy Gull, names that serve to emphasize the distinctiveness of the bird's physical appearance when compared to the many other species of gull. In English the bird became known as Ross's Rosy Gull and, latterly, Ross's Gull.

The scientific naming of Ross's Gull raises some points of particular interest. First, as one of the two described specimens was named twice, it is actually the type (the definition) of two different scientific names, and remains so despite the fact that one name is not valid (i.e., is not accepted as a name for a species). The second point relates to the discovery of this new species and the importance of naming in the development of cultural histories. In the Vienna Natural History Museum is a specimen of the same species of gull that had been collected several years before 1823. This bird was labeled as *Larus collaris* by the museum curator and was in the museum collection by 1818. Had the name of the Viennese bird been properly published with a description, the narrative of the MacGillivray and Richardson story would be incidental and the species' cultural history would have been entirely different, with different nationalistic affinities. The "Vienna" bird was collected in Greenland, probably in May

1813, by the mineralogist Karl Giesecke. He wrote in his diary how "a small gull (called kejukik, kejukingoak) otherwise rare here, was found this spring in quantity. They do not nest in the country." The name "kejukingoak" appears more than thirty years earlier in Otto Fabricius's *Fauna Groenlandica* (1780) alongside yet another scientific name, *Larus cinerarius*, which fell out of usage as it was not clear to what type of bird that name's author, Carl Linneaus himself, was referring. Fabricius described a gull with a dark patch behind its eyes, a description that applies to the species we now call Ross's Gull.[14] The vernacular names suggest that the bird was familiar enough to indigenous people for it to be distinguishable from other seabirds and that it thus formed part of their culture. These episodes serve to demonstrate how a species' cultural history does not necessarily begin with its "regular" scientific discovery and presentation.

## Bodily Transformations: From Corpses to Specimens

In order for Ross's Gull to become a reality for Western science, dead birds needed to be transformed in terms of their physicality and location. Birds are comparatively complicated to preserve, and there was no widely known preservative method until the early nineteenth century, when the recipe for "arsenical soap" was published.[15] However, bird taxidermy has a longer history than previously thought, extending back to the Middle Ages, and leading ornithologists probably had sound knowledge of bird taxidermy although few of their specimens have survived.[16] Early collections contained mounted birds with skins modeled around artificial bodies that were mounted in lifelike positions on bases, with both wire supports through the legs and glass eyes. Most of these early (pre-nineteenth-century) collections have deteriorated to nothing. Birds were preserved as "study skins" from the mid-nineteenth century onward. These had a longer lifespan, being protected from insects and the elements in wooden drawers, but this method ensured that scientific collections became distinct from display collections, lying beyond the gaze of most visitors to museums.[17] The neat (or not so neat) study skins that lie in drawers in museums disguise the extent of their bodily transformation by preserving the outward appearance of the bird. Bloody and oily (and often smelly) corpses had to be transformed into specimens through skill, craft, and tradition. These preserved and prepared objects can, in turn, tell us a great deal about the living thing that they once constituted, but they also have limitations as much of the original specimen has been lost—the body is often merely thrown away (many were probably also eaten). Many museum specimens have been remodeled a

FIG. 5.1 Photograph taken by S. A. Buturlin and marked by him as showing "a room, where we prepared bird skins. Near the window four skins of adult Rhodost. Rosea [Ross's Gull], and four of their pulli." Note the stuffing materials hanging from the ceiling, the trap on the wall, spirits in bottles, and a variety of prepared birds and fresh corpses. Photo © The Manchester Museum.

number of times: both of the Ross's Gulls collected in 1823 have been transformed from skins to mounts and then to study skins, which are considered more "scientific" (see figs. 5.1 and 5.2).

John Murdoch, an American naturalist on the International Polar Year expedition to Point Barrow (1881–83), gives a good description of the challenges of preparing specimens (his first specimen of Ross's Gull was eaten by "Eskimo dogs"):

> Arctic taxidermy has its drawbacks. The carpenter's shop, where I had to work, would not warm up in spite of the little Sibley stove in it, and by the time I had a skin turned inside out and the skull cleaned, the skin would be so stiff from freezing that it would not turn back, and I used to have to warm it at the stove before I could finish the skin. Besides the metal top, which our commanding officer thought was such a neat and cleanly thing to put on my skinning table, used to become uncomfortably numbing to the fingers.[18]

FIG. 5.2 Study skin of Ross's Gull collected during the International Polar Year expedition to Point Barrow. Photo © The Manchester Museum.

Field collectors would often skin a bird and dry it or salt it, to be finished later; other collectors would complete the job themselves in the field. Fridtjof Nansen shot and prepared eight specimens of the bird in 1894 when he was on his expedition attempting to drift over the North Pole, frozen into the ice, and then traveling by sled to Franz Josef Land. Seven of the specimens were prepared as study skins and their bodies preserved in alcohol. The eighth bird, which was headless (he shot its head off accidentally), was also preserved in alcohol. Nansen presented the whole collection to the University of Christiania (Oslo).[19] One of the birds, still in the Oslo Zoological Museum, bears Nansen's original field label as well as a neater museum label; two of the first three birds that Nansen encountered are still in museums, in Bergen and in Cambridge (U.K.). Raymond Newcomb, naturalist on the ship *Jeannette*, shot a number of the rare gulls in 1879 when his vessel was locked in ice in the Arctic Ocean. The ship sank two years later and many of the crew died as they sought to reach land. Despite these incredible hardships Newcomb managed to retain three of the birds' skins under his shirt. These specimens are still in the collections of the Smithsonian Institution in Washington, D.C.

## Scientific (and Not So Scientific) Specimens

When a specimen of Ross's Gull was found (supposedly) in Britain in 1846/47, the species underwent a change in status to become a "British" bird. This had a real and significant impact on the way it was regarded by British ornithologists and collectors, as British birds were among the most popular subjects for collectors of natural history specimens. The leading London-based collectors were familiar with one another, mixing in a variety of business and leisure settings. They were members of the same elitist natural history societies and competed publicly for possession of desirable specimens in natural history auctions.[20] Specimens of Ross's Gull were not available to collectors through most of the nineteenth century and thus represented a blank in collections. The handful of specimens in museums were not generally on display to the public and would have been very unlikely to have been parted with: a skin from Greenland, donated to the British Museum (Natural History), was described as "one of our greatest treasures" by Richard Bowdler Sharpe, curator of the bird collection.[21] The few specimens in existence were exhibited on a handful of occasions at the fortnightly meetings of the Zoological Society of London by John Gould and Alfred Newton, two of the most prominent ornithologists of the nineteenth century (specimens were exhibited latterly by Henry Seebohm and Henry Dresser). To be able to see these specimens at all was a highly privileged position.

Henry Dresser (1838–1915) was a leading British collector; he began collecting birds' eggs and skins when he was a teenager, and by the end of his collecting career (sixty years later) he had one of the finest private collections in Britain. Writing in 1867, when he was twenty-nine years old, he asked an American correspondent to try to obtain any Ross's Gull specimens he could from the Smithsonian Institution, writing that "I feel the want of them much in my collection."[22] The bird represented a gap. Twenty years later, after having had to borrow skins to describe in his famous *History of the Birds of Europe* (1871–82), he wrote once more to the Smithsonian Institution requesting specimens, again unsuccessfully.[23] After a further five years he eventually met with success,[24] leading him to write to the curator at the Smithsonian, "I cannot tell you how greatly indebted I am for your kindness in the matter. *R. rosea* especially is most welcome as I have so long wished to possess it."[25] Dresser gave the Smithsonian four specimens of birds chiefly from Japan in return for the rare gulls.[26] It had taken him nearly thirty years to obtain skins of Ross's Gull. His continuous efforts to obtain them show just how desirable these specimens could be to collectors.

When Russian ornithologist Sergei Buturlin (1872–1938) discovered the regular breeding grounds of Ross's Gull in Siberia in 1905, Henry Dresser arranged for Buturlin's account of the discovery to be published in Britain in the leading ornithological journal, *Ibis*, produced by the British Ornithologists' Union.

Buturlin was elected as a Foreign Member of the Union following the publication. Dresser also arranged for the sale and exchange of Buturlin's Ross's Gull specimens to European and American collectors and museums.[27] Dresser's access to such privileged information and specimens on the one hand, which came as a result of his business links with America and Russia, and to the highly elitist scientific societies and publishing circles on the other hand, helped propel his own career as an ornithologist. Even early in his collecting career (in the 1870s), Dresser intended to bequeath his collection to a public museum when he was finished with it, and sometimes used this intention as a means of obtaining specimens from other museums.[28] His collections were sold (for a nominal sum) to the Royal Scottish Museum (now the National Museums of Scotland, Edinburgh) in the late 1880s and to the Manchester Museum in 1899 (skins, including two of Ross's Gull) and 1912 (eggs, including ten of Ross's Gull).

Stories about the great value of specimens of Ross's Gull circulated among collectors—mostly originating with those who encountered the same birds or who owned their remains. During the 1881–83 expedition to Point Barrow, John Murdoch and colleagues had secured and prepared more than one hundred specimens of Ross's Gull at a time when there were around fifteen specimens in museums around the world. Murdoch wrote that another naturalist (Elliot Coues) "half seriously took me to task 'for vulgarizing this beautiful bird.'" Murdoch was forbidden from publicizing the number of specimens collected, in case the Smithsonian was overwhelmed with requests, although it is equally likely that the Smithsonian did not want the rare specimens to be devalued.[29] Similarly, the English collector Robert Hay Fenton placed a note with two eggs in his collection that the Smithsonian Institution had offered a reward of £100 for eggs and breeding birds,[30] a story without any factual basis.[31] The German ornithologist Herman Schalow, writing after Nansen's supposed discovery of breeding gulls, wrote, "When will man's foot again tread the dreary wastes of those high latitudes where one of the greatest rarities of northern oology is to be found?"[32]

The Norwegian explorer Roald Amundsen (1872–1928) continued to exaggerate the economic value of the rare birds well into the twentieth century, when writing of birds he himself collected (purchased) during his Arctic explorations, in 1920:

> In the bird collection there are 18 good specimens of the extremely rare ring-billed [Ross's] gull: *Larus rossi* and also 9 of the virtually unknown

tundra gull: *Larus sabinii*. The value of these 27 birds alone cannot be ascertained. Real connoisseurs will probably find that they alone will pay for the expedition. It would not surprise me. . . . Apart from the already mentioned 18—official Ring-billed [Ross's] Gulls—I am also sending a case with 20 of the same. . . . Take them out and hide them immediately.[33]

Amundsen instructed his mistress to have the pink feathers from the twenty birds made into two fans, one for herself and the other for Queen Maud (of Norway), in memory of the expedition's journey through the Northeast Passage. Amundsen's biographer discusses the fate of these birds:

> The ring-billed [Ross's] gulls never arrived in London. When Leon [Amundsen] informed him [Roald] of practical complications, new instructions arrived from Seattle: "If fans impossible divide into two. Give one to the Queen and hang the other one up in Uranienborg, preferably under glass." They would hang there, love's winged messengers, stuffed, under glass, in Uranienborg's empty rooms. They are there today—a few strange birds in a glass case, in Roald Amundsen's museum.[34]

The making, naming, and exhibiting of Ross's Gull demonstrate the way in which new markets were created for natural objects: how living things (and their remains) became commodities, valued for their exclusivity. The neat rows of peaceful-looking birds that lie in museum drawers thus disguise the often intense adventures, experiences, and ambitions that were involved in their collection, preparation, and transportation. These narratives are potential sources of wonderful and fantastic cultural interest that are lost, forgotten, or misunderstood without a proper understanding of history and context.

## The Power of Place: Encountering Ross's Gull in an Arctic Context

In 1823/24, the German artist Caspar David Friedrich painted *Arctic Shipwreck*, popularly known as *The Wreck of the Hope*. Architectural iceforms lurch forward and upward in a stomach-churning scene; the wrecked hull of a ship is carried along with the great ice slabs. This picture typifies imagery relating to the Arctic at that time, of a landscape of monumental forces and allegories. For those viewers suitably primed through texts, paintings such as Friedrich's *Arctic Shipwreck* could induce a powerful effect on viewers. It is within these imagined landscapes that characters such as James Clark Ross were placed in the imaginations of readers and viewers, at the time when Ross participated in some of the most adventurous and well-reported attempts on the Northwest Passage. Davidson discusses how paintings such as *Arctic Shipwreck* contribute to—and represent—what he calls "the idea of north," a place that is always elsewhere but which at the same time exerts a pull, as the pole does on a compass needle.[35] The idea of north differs from place to place, from country to country, from occupation to occupation. Thomas Bewick, the famous engraver, was heavily influenced by writings on birds in the Arctic derived from reports of the expeditions and discussed the region at length in the preface to the sixth (1826) edition of his famous, widely read *A History of British Birds.*[36] Among Bewick's numerous vignettes, many showing ships struggling in rough seas and fantastic landscapes, is one showing a ruined ship in an Arctic icescape, extremely similar to the scene in Friedrich's painting.[37] It is worth noting that Caspar David Friedrich was working on his painting at exactly the same time that the first Ross's Gulls were encountered.

Ideas of Ross's Gull are grounded in ideas about the Arctic as a whole and associations with people involved in Arctic exploration.[38] The Arctic has been the source of great fascination since the beginning of the age of exploration. Among imperial powers, this interest became a national obsession during the nineteenth century, when expeditions were launched in search of the Northwest Passage and the lost Franklin expedition, for reasons of imperial expansion and national pride. Interest in the Northwest Passage was transferred to the North Pole, and a flurry of expeditions were organized by imperial powers and smaller European countries with a tradition of Arctic exploration, both government sponsored and privately backed.

Interpretations of the Arctic in European literature and the visual arts emphasized its exotic features: the ways it was weird, bizarre, different, mysterious, evasive, and elusive.[39] It was a place of mysterious forces, of the aurora borealis and the invisible magnetism that drew the compass needle, a land in which "white men" struggled to survive where indigenous peoples had lived for thousands of years, feasting on whale blubber and making knives from iron meteorites. The poles were even proposed as being the source of the origin of life, having been the first spots on the earth to cool after the formation of the planet.[40] The Arctic was without comparison through the nineteenth century in popular terms until the rise in interest in the Antarctic, after which time the two regions became increasingly confused. The Arctic has thus been highly aestheticized through literature and visual culture since the nineteenth century. If the Arctic was depicted as a wasteland, a blank space that humans pitted their wits and their lives against in the name of progress and human endeavor, in more modern times it has also become the focus of concern due to environmental issues such as ozone depletion, pollution, and climate change.

Through the nineteenth century, a small number of Arctic explorers came into contact with Ross's Gulls. As the birds were rare, these encounters frequently made their way into travel narratives and scientific papers. Information on the species was so meager that every single sighting and specimen could be listed until the end of the nineteenth century.[41] These facts and specimens were hard-won. The most insightful commentary on encounters with Ross's Gull comes from Fridtjof Nansen, who wrote in his diary for August 3, 1894, "Today my longing has at last been satisfied. I have shot Ross's Gull." Whether Nansen had articulated this longing beforehand, or whether the statement originated with the benefit of hindsight, is unknown. Nevertheless, the bird clearly excited a sensation of longing, of yearning and ambition, that was satisfied, at least momentarily, through killing the bird. Despite his remote situation, Nansen wrote at length about his encounter in his diary:

> This elusive, strange and rarely seen inhabitant of the mysterious north, a world to which the imagination alone aspires and of which no one knows its coming and going, is that thing, from the first moment I saw these tracts and my eyes surveyed the lonely plains of ice, I had always hoped to discover. And now it came when I was least expecting it, indeed I was out only briefly on a very prosaic errand. As I sat near a hummock my eyes roamed northwards and spotted a bird glide over the big hummocky rise towards the northwest. At first I thought it was a kittiwake, but soon saw that it resembled more an Arctic skua.

This was more or less repeated in the travel narrative subsequently written by Nansen, which was widely read, although missing the fact that Nansen had been on "a very prosaic errand."[42] As a seaman, Nansen would have been well aware of the usefulness of gulls as guides to land. He and his companion shot many other gulls and birds of different species for food, so the shooting of birds, which stands out to modern readers, was an everyday act to the party. John Murdoch described his first encounters with the birds, as usual with the benefit of hindsight: "As I walked up the beach, several flocks of small graceful Gulls passed me, moving towards the northeast, but out of gunshot. As they whirled in the sunshine, I thought I noticed that some of them were rosy underneath. Could they be the famous Rosy Gulls? As may be fancied, I grew a little excited."[43]

The search for Ross's Gull was neither coordinated nor clearly articulated, and encounters mainly appear to have come through chance. Thus it is unclear to what extent the bird was consciously regarded as a symbol of the Arctic to explorers. The clearest reference to a search comes from Howard Saunders (1835–1907), an English ornithologist who specialized in collecting gulls and terns, and in studying Arctic exploration. At the time of the British Arctic Expedition (1875–76), Saunders wrote, patriotically, that he expected the expedition to find evidence of the breeding grounds of Ross's Gull.[44] Despite Saunders's comments, the expedition was unsuccessful: it failed to reach the Pole, a number of men died of scurvy, and no Ross's Gulls were seen. Prefacing his published list of the birds he encountered, the expedition's naturalist, Henry Feilden, managed a hundred words on where the bird *did not* occur, demonstrating the way imperial science was closing in on the bird, even if only by negative evidence.[45] Similarly, following Nansen's discovery of young Ross's Gulls during the famous *Fram* expedition, Richard Bowdler Sharpe, the leading ornithologist at the British Museum (Natural History), framed his comments in an interesting way: "The interest of research in the arctic regions, as regards birds, always sums itself up into one or two directions to the explorer: 'Be sure to find the egg of the Knot, or Curlew-Sandpiper, or find out Ross's Gull—tell us where it breeds.' Dr. Nansen found Ross's Gull breeding, and that is all ornithologists can ask of an arctic explorer, that he solves one or two of these questions."[46] The species' existence entered it into a set list of questions for imperial science to answer, although the extent to which explorers were aware of the existence of such sets of questions is not always apparent.

## Standardizing: Other Literary and Visual Representations of Ross's Gull

As it was an extremely rare visitor to Europe and North America, Ross's Gull was first encountered by most European naturalists through mediated literary and visual representations rather than by direct personal experience. The development of the standard of "Ross's Gull" is now long forgotten. The bird's scientific names are based on the merest description of one specimen by MacGillivray and a slightly longer description of this and another bird by Richardson. Naturalists subsequently described various other stages of plumage. MacGillivray's and Richardson's descriptions are virtually unknown, contained within highly specialized and technical literature, and they would not be familiar to most ornithologists or naturalists. Standardization of the *idea* of the bird has involved a more complex conglomeration and distillation of facts and narratives.

Seabirds feature prominently in all descriptions of the Arctic, for example in Charles Kingsley's *The Water Babies* and Hermann Melville's hyper-sublime poem "The Berg (A Dream)," and although referring to the Antarctic, most educated readers would have been familiar with the albatross in *The Rime of the Ancient Mariner*. Early descriptions of Ross's Gull were largely text based, as for example F. O. Morris's account in his popular *A History of British Birds*, where he describes the bird's color as being of "a deep tinge of 'rich and rare' peach-blossom, red or rose-colour–borrowed, as it were, in the hyperborean regions, the native places of the bird, from the 'Aurora Borealis,' the 'Northern Light,' which, as if to make up for the brief day, transplants the gleam of the morning to gild the long night of the Arctic year."[47] Morris's description was accompanied by a fair plate that is far less evocative than this purple passage.

The most successful and beautiful representations were produced for the expensive subscription books of the mid- to late nineteenth century. The language of representation was highly refined in these books; the majority of plates in the leading books and journals were produced by a handful of artists who based their illustrations on museum study skins. Imagery was thus highly standardized through unwritten representational conventions. Similar standards continue to dominate field guides, which are primarily intended to aid identification and discrimination from similar species. In John Gould's *Birds of Great Britain*, one of the greatest of all bird books, two summer-plumaged adults are shown alongside the winter-plumaged bird supposedly collected in Yorkshire, an impossible combination that collapses time. The English naturalist Lord Lilford (1833–96) employed artists to paint birds for him (he was an invalid and was largely confined to home); he wrote to the artist Archibald Thorburn, "I am sending you today . . . a skin of Ross's Wedge Tailed Gull lent to me by Professor Newton. In Ross's Gull I should be glad if you could make the wedge-shaped tail as conspicuous as possible, and the breast may be brilliantly rose-coloured. An Arctic-sea scene with cloudless pale-blue sky and broken ice-floe, *not bergs*, will best suit this plate."[48] Artists had to imagine the shape of the repositioned bird (from a study skin that lies on its back in a drawer), basing their illustrations on their own firsthand experiences of similar birds, whether as living birds or as illustrations.

In the *Catalogue of Birds in the British Museum*, Howard Saunders gives a very neutral and unimpassioned account of *Rhodostethia rosea*, with one and a half pages given over to the published references on the bird and details of each sighting.[49] Three quarters of a page is given to descriptions of both the plumage of the adult in summer and winter, and the immature plumage. The description of the range is confined to the handful of observations of birds. Breeding distribution was summed up as "propagation as yet unknown." Such descriptions demonstrate the highly refined practices of description by words rather than pictures that typify scientific description from the later nineteenth century.

Firsthand accounts of the bird(s), from travel narratives, read very differently. They tell the reader about the bird's beauty and often present an apparent immediacy, although most were written long after the encounters, when explorers were back at home. Even more poignant are the descriptions that survive from the writings of explorers who failed in their quests and died on their travels, accounts that were salvaged from the wreckage of these failed Arctic explorations.[50] Travel narratives typically mention the bird's distinctive color, often in very beautiful terms: "scarcely a colour"[51] and "each pair appeared as so many roseate points against the bluish ice."[52] The prospect of a rosy gull, with connotations of hopefulness and good times ahead, set against a background of cold, hostile Arctic indifference would be sure to appeal to Victorian sensibilities of duty and the sublime.

Printing technologies varied greatly in their ability to replicate the birds' subtle coloring. Illustrations in nineteenth-century travel narratives were the least successful. Whereas George de Long described one of the birds that Newcomb obtained as "a most valuable prize and rare beyond calculation," the accompanying woodcut resembles a roughly drawn Herring Gull, an abundant and unremarkable species.[53] Baron Adolf Erik Nordenskiöld (1832–1901), a Finnish-Swedish scientist, explored the northern coast of Asia in 1878/79 and encountered Ross's Gull. In the published account of his voyage, the bird is represented by a poor-quality woodcut illustration of a rather generic seagull.[54]

Two particularly fascinating images are included in Collett and Nansen's account of the birds seen on the *Fram* expedition, a photograph and a color illustration. The photograph (fig. 5.3) shows two of the three birds shot on August 3, 1894, a doctored version of a photograph included in Nansen's *Farthest North* (opp. p. 416) tidied up for more "scientific" presentation. The second image (fig. 5.4) would have been prepared with reference to the study skins Nansen had collected. This illustration is singularly interesting as it shows a young Ross's Gull posed, as usual, in profile in the foreground, sitting on an

FIG. 5.3 Photograph of Ross's Gulls shot by Fridtjof Nansen, doctored for "scientific" presentation. From Collett and Nansen, "An Account of the Birds."

RHODOSTETHIA ROSEA (MACG) 1824.
YOUNG IN FIRST PLUMAGE.

FIG. 5.4 Illustration of Ross's Gulls showing the *Fram* in the background. From Collett and Nansen, "An Account of the Birds."

ice floe; another individual flies behind it. The background shows the *Fram* with the ship's crew on the adjacent ice. This is very unusual: the seemingly "scientific" illustration incorporates an element of adventure by portraying the men of the expedition, locating them in the Arctic and within the same visual narrative as the bird.[55]

One type of representation that is worth drawing attention to is the relative paucity of commentary on the bird's calls, and indeed of bird calls in general in nineteenth-century literature. Standardization of bird calls was more complex than standardizing visual representation, at least for many species, until sound-recording technologies became widely available in the twentieth century. Sergei Buturlin discusses the calls of Ross's Gulls in his articles in the *Ibis* in unusual detail, with cries including "á-wo," "claw, claw, claw," and "kiáw, kiáw" among others.[56] Many other species with simpler or more distinctive calls and songs have long cultural histories, for instance the tradition of reporting the first Cuckoo in spring in the United Kingdom. The cultural histories that develop for species are linked, at least in part, with technologies required for description and reproduction.

## Rediscovering Ross's Gull

This essay set out a caricature of scientific discovery, with the discovery of a new species that is preserved and described, and which accumulates a scientific and cultural history. Throughout the essay I have tried to show that this caricature fails to reflect the complex interactions involved in the production and development of a species' cultural history. The apparent solidity and certainty of the process of scientific naming can disguise the social and technical complexities involved, as is well illustrated through the naming of Ross's Gull. The bird was named more or less successfully several times. Failed attempts to describe the species demonstrate a longer Western cultural history that can be difficult to discover. The complex social practices involved in preserving and transmitting specimens, and in describing and publishing them scientifically, are subject to many vicissitudes that would likely influence the development of cultural histories.

The role of type specimens in biology, and the ways they are understood, has changed over time. The development of the practice of depositing type specimens in museums dates from the early nineteenth century (in Britain at least). Descriptions of new species predating the establishment of institutionalized collections were often based on illustrations of specimens, rather than by reference to specimens themselves which were to be preserved in perpetuity. The many—all members of a particular species—are represented by the one (or a few), just as the one represents the many. However, type specimens are often, as Daston notes, not typical: they cannot represent all life stages of a particular species (egg, chick, juvenile, adult, and winter plumage for Ross's Gull, for example).[57] Type specimens may be the first to come to hand rather than some kind of ideal or perfect specimen; they are imperfect synecdoches. The links between names and things underpin modern understandings of nature; in scientific terms, it is the link between a name and a description, based on a type specimen; or as Daston says, "It is the calibration of species—always incorporated in particular specimens—with the holotype and description that forges the chain of transmission [of knowledge]."[58] The example of Ross's Gull shows how the apparent "fixity" of the idea of the species, its reputation (as opposed to its material reality), is not really derived from the type specimens. These specimens are in many ways obscure, hidden from sight, and in the case of this species can have more than one name at the same time. Any assumption that ideas about nature are rooted in perfectly preserved series of lovingly attended and fully understood specimens is highly problematic: the whereabouts of many type specimens are unknown, as naturalists did not clearly identify the specimens they based descriptions on in former times. Many type specimens are also known or presumed to have been destroyed over the course of time, leaving scientific names with no definition beyond the published description.

A fundamental shift in the role of type specimens relates to their functioning as reference points. The overarching goal of nineteenth-century natural history was to classify nature to create an inventory, achieved through reference to (and production of) type specimens. Modern biology is primarily focused on studying nature in light of an understanding of evolution: types become reference points for understanding a changing world, rather than of a static, fixed nature. Nineteenth-century ornithology was led by museum workers who were more familiar with prepared skins than with living birds. Contemporary birdwatchers and field scientists, on the other hand, rarely prepare skins but make observations of living birds using equipment that is often highly advanced, in order to collect data in ways that could not have been imagined previously (e.g., the sound-recording and analysis discussed above). The status of specimen-based museum work has apparently diminished.[59] Collecting birds and eggs, formerly popular pastimes, are either unpopular or downright illegal. The core "image" or idea of a species is far more likely to come from a representation in a book, a wildlife film, or the Internet rather than from a preserved skin or a firsthand experience of a bird seen from afar: very few people are first introduced to particular species through type specimens. These shifts in importance in different fields of activity have a profound effect on the ways that particular animals can retain or lose their standardized image and reputation.

Ross's Gull has been sought out, collected, valued, and exchanged in various ways, by collectors, museums, and latterly birdwatchers. Nussbaum's discourse on objectification is useful in thinking about how animals are treated upon becoming recognized as a "real thing."[60] Objectification takes place when someone, the objectifier, treats the object as a tool for his or her own purposes, lacking in autonomy, as something which can be exchanged or traded with other objects (of the same type or not) and something whose experiences and feelings (if any) need not be taken into account. These ideas resonate with many of the episodes that have been presented in this essay, where people have collected and traded specimens of the gull, often to satisfy their personal desires.

The writings of "great men" such as Nansen inspired others to think about the bird. The armchair geography enjoyed by thousands of readers—thinking of Lubbock's famous line that "we may sit in our library and yet be in all quarters of the earth"[61]—occupied a significant place for readers, providing information not just on geographical localities but also of "landscapes of the mind." In this context, in the aestheticized Arctic, Ross's Gulls were very much elsewhere.[62] Nansen's evocative passage about shooting his first Ross's Gulls, which would have been widely read at the time, has been repeated often.[63] Nansen, in turn, had been greatly influenced by the exploits of Adolf Erik Nordenskiöld.[64] The "tracts" that Nansen had been so influenced by were far more likely to have been (textual) travel narratives from explorers such as Nordenskiöld than obscure descriptions of type specimens in a scientific journal produced in Edinburgh. Michael Densley, who has written more about Ross's Gull in Britain than anyone else, describes how he went from being a teenager who "had never even heard of Ross's Gull" to becoming completely captivated by the bird as a result of written accounts, being particularly influenced by the writings of another enthusiast, James Fisher, and by a handful of illustrations.[65]

## Conclusion: Thinking with Animals Revisited

Birds are, as Mynott says, "good to think with," playing on Lévi-Strauss's "animals are good to think with."[66] But when we are engaging with animals, what are we engaging with? Our ideas about animals are influenced by a mixture of social, textual, and visual practices that are heavily colored by cultural phenomena relating to geography, imperialism, and nationalism; social practices in natural history; image construction and imagination. Cultural activities have influenced how Ross's Gull is understood, just as the (standardized, idealized) bird has influenced human cultural activities. Mere knowledge of the birds' existence signified a level of proficiency and expertise, proficiency that might lead to (or serve to justify) membership in exclusive social groups. Knowledge of the history of discovery signified further connoisseurship and an awareness of nationalistic programs. Engaging with animals can be just as much a social (human) activity as a human-animal engagement, blurring and confusing any imputed relationships and asking for definitions of who "we" are. This two-way directionality of influence not only blurs the separation between nature and culture but also highlights the existence of a set of interacting disciplinary areas: of nature as nature, nature as culture, nature in culture, and culture in nature.

This essay has sought to tease out the social complexities involved in the development of ideas about animals—in the case of Ross's Gull the blurriness between its "scientific" production and its appearance in popular culture, and the contingencies involved in its discovery, naming, and representation. But what benefit does this approach bring? How does it help us understand human-animal relations? In "Three Little Dinosaurs, or A Sociologist's Nightmare," Bruno Latour's apocryphal sociologist seeks to explore where ideas about animals come from, investigating the development of ideas about a particular, long-extinct dinosaur. Are ideas about it based on the real animal, on the scientific conception of the animal, or on its appearances in popular culture? The sociologist gets horribly tangled in the situation, finding that everything apparently influences the other, and he eventually gives up the investigation.[67] Sismondo has commented on this approach that "adding the dimension of social realities necessarily complicates our accounts of scientists' actions. . . . It creates the task of working out the relationships between these social realities and material ones and between the social realities and scientific knowledge. But adding social realities might help us to work through analytical problems. Bruno Latour's 'three little dinosaurs'. . . become simply a puzzle to be worked through, not a 'sociologist's nightmare.'"[68] Similarly, Lewontin has advocated the importance of differentiating the body of knowledge being built into a field from the sociology of that knowledge within a scientific community, an approach that resonates with discussion about the role of type specimens.[69]

The rarity of species such as Ross's Gull ensures that particularly detailed information may be available for interpretation and analysis, and, just as important, in manageable quantities. They can provide an insight into the mechanics and practices involved in the "production" and development of other, less-documented species, although it should be borne in mind that the balance of "popular" and scientific knowledge may be different between different species. As nature-engagement activities and cultural contexts change over time, celebrity species may come and go, and "long-lived" celebrity species will be understood in different ways. Stories and ideas associated with them may be re-engineered and re-appropriated into more contemporary frames of reference. Nevertheless, rare, desirable species such as Ross's Gull present a rich subject for the investigation of the complexities of human-animal interactions and the development of ideas about animals.

## NOTES

1. Nansen's diary of the *Fram* expedition is held at the National Library, Oslo. The digitized version can be viewed at http://urn.nb.no/URN:NBN:no-nb_digimanus_119821 (accessed June 7, 2012).
2. Del Hoyo, Elliott, and Sargatal, *Handbook of the Birds of the World*, 621.
3. Fisher and Lockley, *Sea-Birds*, 234.
4. Densley, *In Search of Ross's Gull*, xxi.
5. See, for example, Potapov, "Birds and Brave Men in the Arctic North."
6. See http://10000birds.com/welcome-wednesday-the-thrill-of-the-chase.htm (accessed June 8, 2012).
7. See Daston, "Type Specimens and Scientific Memory," for detailed discussion of the development of the type concept as it related to specimens as examples, and on the development of preservation practices for type specimens in museums. Johnson, in "Type-Specimens of Birds as Sources for the History of Ornithology," used the type specimens of birds in the collections of the London Natural History Museum to illustrate trends in naming and preservation.
8. Richardson, *Appendix to Captain Parry's Journal of a Second Voyage*, 360.
9. See, for example, Swainson and Richardson, *Fauna Boreali-Americana*, 427.
10. MacGillivray, "Description, Characters, and Synonyms of the Different Species of the Genus *Larus*," 249.
11. MacGillivray, *British Birds*, 618; Saunders, "On the Larinae or Gulls," 208; Mearns and Mearns, *Biographies for Birdwatchers*, 310; Densley, *In Search of Ross's Gull*, 28.
12. See Farber, *Emergence of Ornithology as a Scientific Discipline*.
13. See, for example, Morris, *History of British Birds*, 135; Gould, *Birds of Great Britain*, plate 63; Seebohm, *History of British Birds*, 305.
14. Hjort, "Early Days of Ross's Gull *Rhodostethia rosea* in Greenland."
15. Farber, "Development of Taxidermy and the History of Ornithology."

16. Schultze-Hagen et al., "Avian Taxidermy in Europe from the Middle Ages to the Renaissance."

17. Wonders, *Habitat Dioramas*; Mearns and Mearns, *Bird Collectors*.

18. Murdoch, "Historical Notice on Ross's Gull."

19. Collett and Nansen, "Account of the Birds," 17.

20. Cole, *Egg Dealers of Great Britain*.

21. Nansen et al., "North Polar Problem," 520.

22. Letter from Henry Dresser to George Boardman, Maine, from London, February 1, 1867, Record Unit 7071, Smithsonian Institution Archives, Washington, D.C.

23. Letter to Robert Ridgway, Smithsonian Institution, from Farnborough, April 11, 1887, Robert Ridgway Letters, Blacker-Wood Collection, Department of Rare Books and Special Collections, McGill University, Montreal.

24. Letter to Robert Ridgway, Smithsonian Institution, from London, July 4, 1892, Robert Ridgway Letters, Blacker-Wood Collection, Department of Rare Books and Special Collections, McGill University.

25. Letter to Robert Ridgway, Smithsonian Institution, from London, September 7, 1892, Robert Ridgway Letters, Blacker-Wood Collection, Department of Rare Books and Special Collections, McGill University.

26. Listed as Smithsonian Institution accession number 25966.

27. McGhie and Logunov, "Discovering the Breeding Grounds of Ross's Gull"; McGhie and Logunov, "Henry Dresser and Sergius Buturlin" (in Russian).

28. Letter to Spencer F. Baird, Smithsonian Institution, from London, September 9, 1870, Smithsonian Institution Archives, Unit 7002.

29. Murdoch, "Historical Notice on Ross's Gull," 152.

30. This note can be found in the catalog of Hay Fenton's collection at the Zoology Museum, Aberdeen University, Scotland.

31. McGhie and Logunov, "Discovering the Breeding Grounds of Ross's Gull."

32. Quoted in Palmer and Widman, "Nansen's Discovery of the Breeding Grounds of the Rosy Gull," 176.

33. Bomann-Larsen, *Roald Amundsen*, 190. Note: Ross's Gull is mistranslated as "ring-billed gull" in the English translation.

34. Ibid., 201.

35. Davidson, *Idea of North*.

36. See Spufford, *I May Be Some Time*.

37. Bewick, *History of British Birds*, 185.

38. See Spufford, *I May Be Some Time*; David, *The Arctic in the British Imagination*; Densley, *In Search of Ross's Gull*; Potapov, "Birds and Brave Men in the Arctic North."

39. See Said, *Orientalism*; David, *The Arctic in the British Imagination*.

40. See, for example, Tristram, "Polar Origin of Life Considered in Its Bearing on the Distribution and Migration of Birds."

41. See, for example, Sclater, "On the Specimens of Ross's Gull"; Saunders, "On the Immature Plumage of *Rhodostethia rosea*"; Saunders, "On the Larinae or Gulls."

42. Nansen, *Farthest North*, 414.

43. Murdoch, "Historical Notice on Ross's Gull," 151.

44. Saunders, "On the Immature Plumage of *Rhodostethia rosea*," 485.

45. Feilden, "List of Birds Observed in Smith Sound and in the Polar Basin During the Arctic Expedition of 1875–76," 402.

46. Nansen et al., "North Polar Problem," 520.

47. Morris, *History of British Birds*, 136.
48. Drewitt, *Lord Lilford*, 166–67.
49. Saunders, "Gaviae."
50. De Long, *Voyage of the Jeannette*, 2:616; Densley, *In Search of Ross's Gull*, 35.
51. Bliss and Newcomb, *Our Lost Explorers*, 282.
52. Buturlin, "Breeding-Grounds of the Rosy Gull," 133–34.
53. De Long, *Voyage of the Jeannette*, 1:151–52.
54. Nordenskiöld, *Voyage of the Vega Round Asia and Europe*, 93.
55. The scene is based on a photograph showing one of the ship's crew looking at the *Fram*. See Nansen, *Farthest North*, 392.
56. Buturlin, "Breeding-Grounds of the Rosy Gull," 134–35.
57. Daston, "Type Specimens and Scientific Memory."
58. Ibid., 182.
59. Johnson, "The Ibis." See also Johnson, "Type-Specimens of Birds as Sources for the History of Ornithology."
60. Nussbaum, "Objectification."
61. Lubbock, *Pleasures of Life*, 63.
62. Marshall Gardner, a leading Hollow Earth proponent, continued to believe that Ross's Gulls would be found breeding in the interior of the earth, entering at the poles, although the birds' breeding grounds had already been found. See Gardner, *Journey to the Earth's Interior*, 154, 189.
63. See, for example, ibid.; Densley, *In Search of Ross's Gull*; Potapov, "Birds and Brave Men in the Arctic North"; McGhie and Logunov, "Breeding Grounds of Ross's Gull."
64. Bomann-Larsen, *Roald Amundsen*, 9.
65. Densley, *In Search of Ross's Gull*, xviii, 2.
66. Mynott, *Birdscapes*, 262.
67. Latour, "Three Little Dinosaurs, or A Sociologist's Nightmare."
68. Sismondo, *Science Without Myth*, 58.
69. Lewontin, "Theoretical Population Genetics in the Evolutionary Synthesis," 60.

## BIBLIOGRAPHY

Bewick, Thomas. *A History of British Birds*. Vol. 2, *The Water Birds*. Mem. ed. Newcastle-upon-Tyne (U.K.): R. Ward and Sons, 1885.
Bliss, Richard W., and Raymond L. Newcomb. *Our Lost Explorers: The Narrative of the Jeannette Arctic Expedition as Related by the Survivors, and in the Records and Last Journals of Lieutenant de Long*. Hartford: American Publishing Co., 1882.
Bomann-Larsen, Tor. *Roald Amundsen*. Translated by Ingrid Christophersen. Stroud (U.K.): Sutton Publishing, 2006. First published in Norwegian: Oslo: Cappelen, 1995.
Buturlin, Sergei A. "The Breeding-Grounds of the Rosy Gull." *Ibis* 48, no. 1 (1906): 131–39.
Cole, Andrew C. *The Egg Dealers of Great Britain*. Horsforth (U.K.): Peregrine Books, 2006.
Collett, Robert, and Fridtjof Nansen. "An Account of the Birds." In *The Norwegian North Polar Expedition, 1893–1896*, edited by Fridtjof Nansen. Scientific Results, vol. 1, no. 4. London: Longmans, Green, and Co., 1899.

Daston, Lorraine. "Type Specimens and Scientific Memory." *Critical Inquiry* 31 (2005): 153–82.

David, Rob. *The Arctic in the British Imagination, 1818–1914.* Manchester: Manchester University Press, 2000.

Davidson, Peter. *The Idea of North.* London: Reaktion Books, 2005.

Del Hoyo, Josep, Andrew Elliott, and Jorgi Sargatal, eds. *Handbook of the Birds of the World.* Vol. 3, *Hoatzin to Auks.* Barcelona: Lynx Edicions, 1996.

De Long, Emma. *The Voyage of the Jeannette: The Ship and Ice Journals of George W. de Long, Lieutenant-Commander U.S.N., and Commander of the Polar Expedition of 1879–81.* 2 vols. Boston: Houghton, Mifflin, and Co., 1884.

Densley, Michael. *In Search of Ross's Gull.* Horsforth (U.K.): Peregrine Books, 1999.

Drewitt, Caroline M. *Lord Lilford: A Memoir by His Sister.* London: Smith, Elder, and Co., 1900.

Farber, Paul L. "The Development of Taxidermy and the History of Ornithology." *Isis* 68, no. 244 (1977): 550–66.

———. *The Emergence of Ornithology as a Scientific Discipline, 1760–1850.* Baltimore: Johns Hopkins University Press, 1997.

Feilden, Henry W. "List of Birds Observed in Smith Sound and in the Polar Basin During the Arctic Expedition of 1875–76." *Ibis* 19, no. 4 (1877): 401–12.

Fisher, James, and Ronald M. Lockley. *Sea-Birds: An Introduction to the Natural History of the Sea-Birds of the North Atlantic.* London: Collins (New Naturalist), 1954.

Gardner, Marshall B. *A Journey to the Earth's Interior; or, Have the Poles Really Been Discovered?* Aurora, Ill.: Published by the author, 1920.

Gould, John. *The Birds of Great Britain.* Vol. 5. London: Privately published, 1873.

Hjort, Christian. "The Early Days of Ross's Gull *Rhodostethia rosea* in Greenland." *Dansk Ornithologisk Forenings Tidsskrift* 79 (1985): 152–53.

Johnson, Kristin. "*The Ibis*: Transformations in a Twentieth-Century British Natural History Journal." *Journal of the History of Biology* 37 (2004): 515–55.

———. "Type-Specimens of Birds as Sources for the History of Ornithology." *Journal of the History of Collections* 17, no. 2 (2005): 173–88.

Latour, Bruno. "Three Little Dinosaurs, or A Sociologist's Nightmare." *Fundamenta Scientiae* 1 (1980): 79–85.

Lewontin, Richard C. "Theoretical Population Genetics in the Evolutionary Synthesis." In *The Evolutionary Synthesis: Perspectives on the Unification of Biology,* edited by Ernst Mayr and William B. Provine, 58–68. Cambridge: Harvard University Press, 1980.

Lubbock, John. *The Pleasures of Life.* London: Macmillan and Co., 1887.

MacGillivray, William. *British Birds.* Vol. 5. London: William S. Orr and Co., 1852.

———. "Description, Characters, and Synonyms of the Different Species of the Genus *Larus.*" *Memoirs of the Wernerian Natural History Society* 5, no. 1 (1824): 247–76.

McGhie, Henry A., and Dmitri V. Logunov. "Discovering the Breeding Grounds of Ross's Gull: 100 Years On." *British Birds* 98, no. 11 (2005): 589–99.

———. "Henry Dresser and Sergius Buturlin: Friends and Colleagues" [in Russian]. In *Materials for the Second International Conference Dedicated to the Memory of Sergei Buturlin, Held in Ulyanovsk 21.09–24.09.2005,* edited by Olga Borodina et al., 40–53. Ulyanovsk: Korporatsiya tekhnologii prodvizheniya, 2006.

Mearns, Barbara, and Richard Mearns. *Biographies for Birdwatchers: The Lives of Those Commemorated in Palearctic Birds' Names.* London: Academic Press, 1988.

———. *The Bird Collectors.* London: Academic Press, 1998.

Morris, Francis O. *A History of British Birds.* Vol. 6. London: Groombridge and Sons, 1857.

Murdoch, John. "An Historical Notice on Ross's Gull." *Auk* 16, no. 2 (1899): 146–55.

Mynott, Jeremy. *Birdscapes: Birds in Our Imagination and Experience.* Princeton: Princeton University Press, 2009.

Nansen, Fridtjof. *Farthest North.* Westminster: Archibald Constable and Co., 1897.

Nansen, Fridtjof, Joseph Hooker, Henry W. Feilden, and George Nares. "The North Polar Problem: Discussion." *Geographical Journal* 9, no. 5 (1897): 505–28.

Nordenskiöld, Adolf Erik. *The Voyage of the Vega Round Asia and Europe, with a Historical Review of Previous Journeys Along the North Coast of the Old World.* London: Macmillan and Co., 1882.

Nussbaum, Martha C. "Objectification." *Philosophy and Public Affairs* 24, no. 4 (1995): 249–91.

Palmer, Theodore S., and Otto Widman. "Nansen's Discovery of the Breeding Grounds of the Rosy Gull." *Science,* n.s., 5, no. 109 (1897): 175–76.

Potapov, Eugene R. "Birds and Brave Men in the Arctic North." *Birds International* 2, no. 3 (1990): 73–83.

Richardson, John. *Appendix to Captain Parry's Journal of a Second Voyage for the Discovery of a North-West Passage from the Atlantic to the Pacific Performed in His Majesty's Ships Fury and Hecla, in the Years 1821–22–23.* London: John Murray, 1825.

Said, Edward. *Orientalism: Western Conceptions of the Orient.* London: Penguin, 1991 [1978].

Saunders, Howard. "Gaviae (Terns, Gulls and Skuas)." In *Catalogue of the Birds in the British Museum,* vol. 25, *Gaviae and Tubinares,* edited by Howard Saunders and Osbert Salvin, 1–339. London: British Museum Trustees, 1896.

———. "On the Immature Plumage of *Rhodostethia rosea.*" *Ibis* 17, no. 4 (1875): 484–87.

———. "On the Larinae or Gulls." *Proceedings of the Zoological Society of London,* pt. 1 (1878): 155–211.

Schultze-Hagen, Karl, Frank Steinheimer, Ragnar Kinzelbach, and Christoph Gasser. "Avian Taxidermy in Europe from the Middle Ages to the Renaissance." *Journal für Ornithologie* 144, no. 4 (2004): 459–78.

Sclater, Philip L. "On the Specimens of Ross's Gull." *Ibis* 7, no. 1 (1865): 103–4.

Seebohm, Henry. *A History of British Birds, with Coloured Illustrations of Their Eggs.* Vol. 3. London: Porter and Dulau and Co., 1885.

Sismondo, Sergio. *Science Without Myth: On Constructions, Reality, and Social Knowledge.* Albany: SUNY Press, 1996.

Spufford, Francis. *I May Be Some Time.* Pbk. ed. London: Faber and Faber, 1997.

Swainson, William, and John Richardson. *Fauna Boreali-Americana; or, The Zoology of the Northern Parts of British America, Containing Descriptions of the Objects of Natural History Collected on the Late Northern Land Expeditions Under Command of Captain Sir John Franklin.* London: John Murray, 1831.

Tristram, Henry B. "The Polar Origin of Life Considered in Its Bearing on the Distribution and Migration of Birds." *Ibis* 29, no. 2 (1887): 236–42; 30, no. 2 (1888): 204–16.

Wonders, Karen. *Habitat Dioramas: Illusions of Wilderness in Museums of Natural History.* Uppsala: Uppsala University, 1993.

**6**

## A Dog of Myth and Matter: Barry the Saint Bernard in Bern

*Liv Emma Thorsen*

The most famous Saint Bernard in history, a rescue dog named Barry, has been on display in the Natural History Museum of Bern since 1815. The dog died in 1814, at fourteen years of age. He became a legend during his own lifetime, and is still remembered because his mortal frame has been preserved. Of equal—or perhaps even greater—importance, his canine spirit has been kept alive in legends, poems, pictures, and, notably, through the museum's own memory work of Barry. At the same time, the dog has been a tactile thing that has sustained and given evidence to his mythical biography. The stuffed Barry is an example of material objects that invite us to talk about what they are in themselves, as well as what they mean. As science historian Lorraine Daston states, certain things may "helpfully epitomize and concentrate complex relationships that cohere without being logical in the strict sense."[1] Barry's biography serves as a good example of the shifting ways by which society and culture influence representations of animals in natural history museums. It also highlights the importance of taxidermy in directing our gaze and thus molding our conceptions of nature and the natural. Stuffed animals, on the one hand, have some authenticity and materiality by definition, but they also create and satisfy preconceived notions of the cultural animal.

I thank Marc Nussbaumer at the Natural History Museum of Bern for his patient assistance, and Henry McGhie, Karen Rader, and Adam Dodd for useful comments and linguistic corrections. All German translations have been provided by the author.

Barry will be followed through two fields of study: that of myth and imagination, and that of nature and beastliness. The two fields are unified in the stuffed dog and in the history of the mount itself. If Barry had been an anonymous Alp dog, his stuffed skin would probably have been discarded long ago. Because of his fame, however, his skin has been kept, restored, transformed, and reinterpreted. To pursue the double position of this dog, I will draw an outline of what I call "Barrylore" and then present what can be surmised as Barry's natural history. The nineteenth century's conception of the faithful dog and the construct of the Saint Bernard breed are also pieces in Barry's history as a principal character, in written and pictorial narratives, and as a museum item. In this context, it is useful to point to ways of writing about and collecting nature in the beginning of the nineteenth century, to explain why the remains of a domesticated dog were preserved in a naturalia cabinet. Most Western natural history museums contain very few domestic specimens, and where they *are* kept, these are rarely displayed today. My main concern is thus to examine the many different ideals and representations a single animal can stand for, and to demonstrate, like the Ross's Gull in Henry McGhie's chapter in this volume, how those representations can guide cultural and scientific appliances.

## Hospice, Monks, "Marroniers," and Dogs

Barry's vita and fame must be seen against the broader history of the Hospice du Grand-Saint-Bernard, one of the three most famous hospices in the Middle Ages, built on the initiative of the Catholic Church to assist, house, and protect pilgrims. Since the middle of the eleventh century, travelers who crossed the Great Saint Bernard Pass could rest and eat safely in the hospice, founded by the archdeacon of Aosta in Piedmont, Bernard of Menthon, later known as Saint Bernard. The pass has a maximum elevation of 8,100 feet (2,469 meters) and has been used since Roman times by travelers through the western Alps between the Swiss Martigny in Wallis and the Italian Aosta. The hospice is situated on the Swiss side of the Swiss-Italian border. Great Saint Bernard is the lowest pass on the ridge containing the two highest summits in the Alps, Mont Blanc and Monte Rosa. The hospice was placed under the care of Augustinian monks, whose main task was to secure a safe passage for travelers. The Swiss naturalist and alpinist Karl Friedrich August Meisner describes the pass as "naked, barren and surrounded by peaks covered with eternal snow."[2] In 1484, the Holy Chair settled the details regulating a rescue service. Pilgrims, poor people, and others who climbed the mountains were to be provided with clothes, food, and whatever else they needed. Between November 11 and May 4, guides called "marroniers" were instructed to leave the hospice daily to patrol the paths both southward to Italy and northward to the canton of Wallis. The marroniers were laypersons in service of the hospice. Travelers were to be offered bread and wine in the hospice and provided with shelter when the weather was bad. When they left, they were to be accompanied by a marronier. Travelers who only passed by the hospice could request bread, wine, and bouillon.[3]

In the nineteenth century, the hospice became particularly famous for its dogs. This fame was mainly due to their achievements as avalanche dogs. The dogs had been newcomers to the hospice compared to their human cohabitants. Precisely when the monks began keeping dogs in the hospice, or where the dogs came from, is not known. One presumes that the ancestors of the famous Saint Bernard breed were brought to the hospice in the second part of the seventeenth century.[4] At the end of the eighteenth century, dogs the size of a modern Bernese mountain dog were kept in the hospice.[5] Barry's subsequent fame was based on his work as a rescue dog, but the dogs' main roles were as watchdogs and as pathfinders. Dogs also worked in the hospice's kitchen as turnspits. People of all kinds met in the desolated hospice, and the heavy-built dogs, usually kept in the entrance hall, pacified quarrelers and guarded the hospice's possessions. The dogs' ability to scent and follow a path was particularly valuable during wintertime: a dog could scent a path hidden under a meter of snow, while his stout body worked as a plow through the snow and made the passage safer for his follower. When visibility was poor, the marronier navigated by holding onto the dog's tail, which floated over the snow. The close cooperation between human and dog paved the way for training the canines as rescue dogs.[6] The canine era of the hospice lasted about 250 years, during which time between 2,000 and 2,500 people were saved from perishing in the mountains.[7]

Authors writing about the hospice dogs in the first half of the nineteenth century attributed their descent to Tibetan mastiffs and Roman fighting and war dogs, as well as to crosses between the Great Dane and local sheep dogs.[8] Marc Nussbaumer, an expert on Swiss canine skull morphology, maintains that it is more likely that the hospice dogs were "Sennenhünde," a branch of the bigger dogs kept by the peasants in the adjacent valleys. Over time, hospice dogs interbred with local canines from neighboring parishes.[9] The dogs were variable in appearance, in terms of coat texture, colors, and markings.

Puppies could be born with short as well as long coats; those with long coats were useless as rescue dogs because a dog working in snow needed a dense, short coat with well-developed underwool. Long-coated puppies were given away. The dogs' skulls varied in shape; some dogs had a short muzzle and a well-defined "stop" between muzzle and forehead, while other dogs lacked the stop. (The stop is "the degree of angle change between the skull and the nasal bone near the eyes.")[10]

Barry must have distinguished himself during his own lifetime. Born in 1800, he served as a rescue dog for twelve years, an unusually long service for a dog that had to endure tough work under extreme conditions. Several dogs were known to have been lost or killed on duty or to have suffered from rheumatism and other bodily weaknesses. Barry lived another two years in Bern before he died in 1814. Further evidence of his singularity comes from the monks' practice of naming the best male dog in the pack Barry.[11] A description of Barry's character and abilities, and the reason why he was stuffed for display in the museum in Bern, is given by Meisner in the yearbook *Alpenrosen* in 1816:

> For twelve years Barry was tireless and faithful in his service for the victims, and he alone has throughout his life saved more than forty persons from death. The eagerness he demonstrated through his deeds was extraordinary. He never needed to be admonished in this service, nothing could keep him in the monastery as soon as the sky became clouded, fog appeared and blizzards announced themselves from a distance; from that point on he would restlessly and barking range around, and he would not tire, again and again returning to the dangerous places, whether or not he could prevent anyone from sinking down into the snow, or dig up one of those already buried under the snow. And if he could not help himself, he would in long leaps rush back to the monastery and try to get help.
>
> When the noble faithful animal grew old and weak, the honorable prior of the monastery sent him with his servant to Bern requesting that after Barry's death that followed in 1814, he should be displayed in our museum. "It is," this sensitive man wrote, "pleasant and at the same time a comfort for me to think that this faithful dog that has saved the lives of so many people, will not be soon forgotten after his death."[12]

Meisner explains why the dog was naturalized, and provides the reader with the canine's principal virtues: faithfulness, courage, and intelligence. As I will show, with the stuffed Barry as material evidence of the historic dog, the mythical dog could float freely.

## A Specimen for His Time

For the up-and-coming bourgeoisie and, to a lesser degree, common people, Barry's life span ran parallel to a common interest in natural history. Naturalists, scholars, and enthusiastic amateurs or *Liebhaber* collected and exchanged all manner of natural objects, including mammals, birds, shells, fossils, insects, fish and crustaceans, minerals, and plants. Old museums expanded their collections, and new collections were established. Being a naturalist in this period was to be a jack-of-all-trades: someone who knew how to combine knowledge of nature with skill in collection and conservation practices, as well as the art of writing for a broad circle of readers. Collecting and describing nature were intertwined activities for many naturalists.[13] One good example is Karl Friedrich August Meisner. Meisner was appointed professor of natural history and geography in 1805, and he corresponded with scientists outside Switzerland. He roamed the mountains and collected, observed, and drew from nature, writing several books on natural history. From 1801 onward he classified and organized the collections in the naturalia cabinet in Bern, in which Barry later was displayed.[14]

By the nineteenth century, natural history books were no longer written in Latin, but rather in the author's native language.[15] Many of these were manuals on national fauna, spiced with anecdotes and stories about the actual animal. Christoph Irmscher designates the poetics of natural history before Darwin as "located in the cross-roads of the Linnaean taxonomy and *belles lettres*, wavering between the demands of precise description and the seductions of the narrative."[16] Another feature that distinguishes the mediation of natural history in the beginning of the nineteenth century from current practice was the inclusion of domestic animals in natural history books and collections. Heinrich Rudolph Schinz, whose career was very much like that of Meisner's, included the domestic dog in *Naturgeschicte der in der Schweiz einheimschen Säugetiere* (Natural history of Swiss mammals), published together with Jakob Römer in 1809 and addressed to "Kenner und Liebhaber," as well as in his *Fauna Helvetica*, published thirty years later. Schinz gives prominence to the dogs in the hospice at Great Saint Bernard.

The 1809 entry on the dog contains fifteen pages and is by far the most extensive, exceeding entries on other domesticated animals such as horses, sheep, and even the emblematic animal of the Alps, the ibex. Here the lifesaving hospice dogs are contrasted with the man-hunting bloodhounds of Cuba and St. Domingo: "Nowhere, as far as we know, are they [dogs] used in a purpose more friendly to man, and this fact belongs exclusively to the history of the Swiss dog."[17] The entry on the dog is much shorter in 1839, but in this text Barry's deeds resonate. He is presented as the last descendant of the pure breed in the hospice, a dog that "saved forty people's lives and is displayed in the museum in Bern."[18] These statements suggest two reasons why Barry became emblematic: one that embeds him in Swiss cultural history, and a second that inserts him in Swiss natural history. Being the most outstanding among the best of canines, and being a Swiss dog, Barry was a perfect specimen for his time.

## Faithful

As Brita Brenna demonstrates in chapter 2, the glass case gives museum items their import and authority.[19] A glass case also framed Barry. A picture of an early display of Barry shows the dog in a cubical glass case, carrying a pyramidal vitrine divided in three sections, each section containing two small mammals, all crowned by a perch supporting an owl with outspread wings (fig. 6.1).[20] Barry must have been considered the main piece in the composition, since a small picture of him, with his name underneath, was placed on top of the glass case. He looks old and tired, but the physical arrangement of the specimens presents the dog as the featured animal. This preference corresponds to cultural conceptions of the dog in the nineteenth century, when the dog was coined man's most faithful companion. French historian Alfred Franklin proclaimed that the dog ranked number two after the human because of its emotional life and developed intelligence.[21] The prominent English veterinarian William Youatt introduced his book *The Dog* with a devoted declaration: "The Dog, next to the human being, ranks highest in the scale of intelligence, and was evidently designed to be the companion and the friend of man."[22] Among all domesticated animals, the dog alone demonstrated unselfish devotedness. So sublime is the dog that "if any of the lower animals bear about them the impress of the Divine hand, it is found in the dog."[23] Not only was the dog a noble creature, but its natural abilities of swiftness and strength and its eminent sense of smell had probably been integral to the establishment of human societies.[24]

The dog's unfaltering loyalty held a strong appeal for nineteenth-century artists. English art critic Philip Gilbert Hamerton wrote in 1874 that "although

FIG. 6.1 The oldest picture of the mounted Barry in a glass case, 1883. From Barbou, *Le chien*, 107.

dogs have been more or less painted and carved since men used brush and chisel, they have never held so important a position in art as they do now. The modern love of incident in pictures, the modern delight in what has been aptly called 'literary interest' as distinguished from the pure pleasure of the eyes, naturally induce us to give a very high place to dogs, which more than all other animals are capable of awakening an interest of this kind."[25] A century later, many nineteenth-century animal paintings would be labeled vulgar or kitsch. Art historian Kenneth Clark states that art conveys many good canine portrayals, but unfortunately sentimentality invaded animal portraiture in the nineteenth century, often expressed through vulgar anthropomorphism, and "the chief victims were dogs."[26] He also maintains that dogs have fueled bad poetry, while cats often have played important parts in literature.[27] However, to publish dog poetry seemed to be an obscure business even in the canine's heyday. Knowing that a collection of British dog poems would be ridiculous to many, in his preface to *The Dog in British Poetry* published in 1893, Robert Maynard Leonard legitimized the book by quoting Bloomfield: "An honest dog's a nobler theme by far than many a one chosen by poets."[28] Barry figures in two of the poems in this volume. Samuel Rogers, the author of "Barry, the Saint Bernard," traveled in 1814 through Switzerland to Italy and might have heard about the dog during this trip.[29] Barry's conduct is described "as guided by a voice from Heaven" when he finds and digs a man out from the snow. The poem "The Dog of Saint Bernard's," by the religious educational author Caroline Fry, is a morality tale in which the dog saves a traveler, but also frightens him; the traveler then kills the Saint Bernard that was his savior.

While eighteenth-century dog lovers described their furry friends with humor and friendly irony, even in elegies, the relationship between the master and his dog was embraced with sincerity in the next century. The majority of the elegiac poems to be found in *The Dog in British Poetry* were written by authors born late in the eighteenth century and early in the nineteenth, and they indicate a stronger emphasis on the emotional bond between people and their dogs. Perhaps the most famous poem in English literature mourning a dead dog is Lord Byron's "Boatswain, His One Friend," an elegy to his New-foundland dog that died of rabies in 1808. The dog was buried on Newstead Abbey, and on the funereal monument is inscribed the following:

> Near this Spot
> are deposited the Remains of one
> who possessed Beauty without Vanity.

Strength without Insolence.
Courage without Ferosity,
and all the virtues of Man without his Vices.
This praise, which would be unmeaning Flattery
if inscribed over human Ashes,
is but a just tribute to the Memory of
BOATSWAIN, a DOG
who was born in Newfoundland, May 1803
and died at Newstead Abbey, Nov.r 18th. 1808.[30]

In the conclusive last stanzas of the poem that follows, Byron addresses the stranger who might contemplate the monument:

Pass on—it honours none you wish to mourn:
To mark a friend's remains these stones arise;
I knew but one, and here he lies.

Dog literature of the period demonstrates that although the dog could be elevated to a morally superior creature when compared to a human, it was also presented as the archetype of the loyal servant. A dog was a canine friend in service of humans, sometimes a heroic friend who sacrificed its life to save its master. English natural history writer and teller of morality tales Rev. John G. Wood's works *Stories and Anecdotes of Dogs* (1856) and *Man and Beast Here and Thereafter* (1875) as well as dog lover George J. Jesse's *Researches into the History of the British Dog* (1866), among others, offered heart-wrenching, sentimental anecdotes about loyal and courageous dogs. The faithful dog featured in poems and stories and on canvases was often a poor man's dog, thus facilitating its way into the human heart and mind.[31] The most popular themes presenting the faithful dog were "the incredible journey," "the incredible waiting time," "the incredible rescue," and "the mourning dog."

Prominent hero dogs of the nineteenth century were the Saint Bernard and the Newfoundland. Their fame was nourished by the image of a big, strong dog in service of man, whether he was out in the wild and capricious ocean or threatened by avalanches when climbing the steep and slippery mountains. Barry fit easily into the latter narrative. The stories present him as a faithful servant of humans from as early as 1816, in Meisner's short text. The prior's intention in having Barry stuffed and displayed was, according to Meisner, to ensure that the dog's loyalty should be remembered after its death. But the good prior could hardly have imagined how widely and vividly Barry's fame would spread.

## "Der heilige auf dem St. Bernhard": The Holy One of Saint Bernard

To come to a better understanding of Barry's enduring fame, it is important to consider that the Saint Bernard became a very popular breed in the second half of the nineteenth century, when it expanded from being considered only as a rough working breed to a breed launched as a "family dog." Dog manuals demonstrate that breeds are not only constructed through breeding, but also authorized by a mythology that explains each particular breed's history and descent. They are biology as well as culture. As Edward Ash states, "Every Saint Bernard was a hero—perhaps untried, but a hero all the same."[32] In her book on the breed, W. F. Barazetti emphasizes both the individuality of the Saint Bernard and the close connection between the breed and Christianity, making them "so much more than ordinary dogs." The Saint Bernard does not take orders, being a friend rather than a slave. The dog's character is selfless-ness, and it is so intelligent that it can act on its own in order to save and assist distressed people: "The St. Bernard has been bred to a Christian purpose, to be a truly good Samaritan."[33]

Two persistent anecdotes highlight Barry's extraordinary abilities as a rescue dog and his devotedness to humans: "the rescue of the child" and "Barry and the traveler." They have been repeated in different languages during the nine-teenth and early twentieth centuries in dog manuals and in educational books for juvenile readers. In "the rescue of the child," Barry becomes a superhero dog for the first time in a story that was recounted in Swiss professor Peter Scheitlin's book *Versuch einer vollständigen Thierseelenkunde* (1840). According to Swiss geologist and cynologist Albert Heim, the most imaginative narra-tives about Barry can be related to Scheitlin's book on the souls of animals.[34] Scheitlin had met Barry in 1812 when the dog still lived in the hospice—Barry had in fact growled at Scheitlin.[35] The old dog's aggression cannot have affected his admirer, however, who many years later wrote confidently, "If I had been a miserable man, you would not have snarled at me."[36] Barry was a perfect animal for a naturalist who was collecting examples to demonstrate that animals have spiritual gifts.

To Scheitlin, Barry was "Der heilige auf dem St. Bernhard," the Holy One of Saint Bernard, and thus the most admirable dog known in history. As he directly addressed the dog in his eulogy, "You were a big, profound man-dog with a kind soul for the unhappy."[37] If Barry had been born a human being, what could he have achieved? He would have founded hundreds of holy orders and monasteries.[38] Indeed, Scheitlin's enthusiasm pushed him so far as to present Barry as a savior in canine form, declaring, "You are the opposite of a sexton. You render the dead their resurrection."[39] This sentence introduces Barry's most famous deed: saving a little boy from perishing in the snow. Barry digs him out of the snow, revives him, and in a friendly way invites the boy to sit up on his back. Riding on the dog, the child is carried safely to the hospice. This anecdote seems to predate the historic Barry, because a very similar version was printed in Anne Francois Joachim Frévilles's book *Histoire des chiens célèlebres, entre-mêlée de notices curieuses sur l'historie naturelle, pour donner le gout de la lecture à la jeunesse* (1796), published in German in 1797.[40]

Scheitlin's portrayal of Barry is an example of how Barry was transformed from being a clever working dog to serving as a canine ideal for humans, and it must have offered fertile soil for the many versions of the anecdotes that grew around the proper dog. In some later versions the child is a boy, in others a little girl. When Paris got its first dog cemetery, Le Cimitière des chiens d'Asnières-sur-Seine in 1899, it was rounded off with an impressive entrance gate, a reception room, a crematorium, and a park with a central monument: a memorial stone showing Barry with a little girl sitting on his back. On the monument we read, "Il sauva la vie à 40 personnes . . . Il fut tué par la 41ème!" (He saved the lives of forty people . . . He was killed by the forty-first!). This claim that Barry rescued forty people is common. The number forty has a specific significance in Christian culture and is repeated in several biblical incidents. To ascribe to the dog the virtue of having saved forty humans reinforced his aura as an almost celestial creature. The fact that the monks did not register the individual dog's doings and deeds has not enfeebled the magic of the number.[41]

The historic, peaceful death of the most famous Saint Bernard seems to have been passed over by his most ardent admirers. A proper hero has to encounter a violent and tragic death. In the anecdote "Barry and the traveler," brave Barry tries to rescue a man, but the man mistakes the dog for a wolf and fatally stabs him. Sometimes the confused traveler is attributed a nationality, sometimes he is an anonymous stranger. L. M. Leonard branded this story "a vulgar mistake" and points at Longfellow's poem "Excelsior," in which the faithful and anonymous dog from the hospice finds the young traveler dead.[42] Unsurprisingly, the "vulgar" version became the most popular, and inspired authors of poetry and prose to mix fantasy and fact in creative ways. In one version, the badly injured dog is brought to Bern; in a second, the bleeding Barry dragged himself back to the hospice, and his malefactor is found and saved by following the blood track. That Barry sacrifices his blood to save a human being associates the lore of the dog with another religious connotation.

Nationalistic elements were also added. In a story written by Italian Carlo Alberto Girardin, Barry and a mountain guide accompanied a Sardinian messenger traveling to his family in the village San Pietro.[43] An avalanche killed and buried the two men, and their bodies were found in the following spring after the snow had melted. This story should, according to the author, date from 1800, the year Napoleon defeated the Austrians in the Battle of Marengo, the very year the historic Barry was born. Barry's death in 1814 was an insignificant event, in the year that saw the end of the Napoleonic Wars. Conversely, Barry's fame as a faithful and clever dog in the service of the suffering must have appealed to people's imagination in the post-Napoleonic period, when Europe was torn between reactionism and radicalism. In fact, Napoleon himself had arrived at the hospice on May 20, 1800, with forty-six thousand soldiers and seven thousand horses on his way to Piedmont, where he conquered the Austrians in the Battle of Marengo three weeks later.[44] Napoleon's assault on Piedmont is echoed in a Danish book on the Saint Bernard breed. The story is much elaborated and with direct speech, and evidently the French emperor offered the female author the opportunity to speak for democracy. Here, a sequence is added to Barry's last heroic deed: when Barry was about to sacrifice his life for freedom, he survived the ten deep cuts caused by the young son of a free peasant, who was fleeing to prevent enlistment in "power-seeking emperor Napoleon's [army]."[45]

A dramatic end renders stories about the faithful dog especially moving. In *A Book of Famous Dogs* (1937) Albert Payson Terhune weaves elements from many of the different versions mentioned above. Barry is retired at ten years of age, but five years later he rushes out to save a soldier who has lost his way; the soldier stabs the dog to death because he thinks Barry is a wild beast.[46] However, the most imaginative statement on the connection between Barry's historical identity and the mounted dog has been given by Kate Sanborn. To her, the fact that the stuffed dog was on display in the museum did not rally her fantasy. In *Educated Dogs of Today* (1916) Sanborn presents Barry as Saint Bernard's own dog. The dog that lived nearly one thousand years ago is still to be seen, and is truly, to quote Sanborn, "a triumph of the taxidermist's art"![47]

The many versions of the two narratives demonstrate the potential of imagery and creativity inherent in a dramatic anecdote with a big, loyal dog in the starring role. The stories are tied to geographic places and related to historic events, albeit blurred and adjusted to actual ideas and cultural preferences. The founders of the dog cemetery in Asnières portrayed Barry with the child as the essence of canine fidelity. When the management of Hartsdale Pet Cemetery in New York, flagged in advertisements as "America's most prestigious pet cemetery," erected a central monument in 1923, they choose to memorize the war dogs in World War I with the form of a bronze German Shepherd in the service of the Red Cross.[48] The hero dog—usually a male—was still meaningful enough to be focused on, but he was no longer a romantic, individual canine rescuing people from nature's forces; rather, he was one of thousands of anonymous quadrupeds that had been enrolled to serve in a manmade inferno of trenches and battlefields.

## Stuffed, Mounted, and Drawn

The arrival of a new millennium was celebrated all over the world, but one rather peculiar event occurred in Bern, where the Natural History Museum celebrated Barry's two hundredth anniversary with a special exhibit, *Barry: Eine Hommage an die Nase* (Barry: A tribute to the nose). The arguments for doing so were twofold: first, that Barry was one of the most important attractions in the museum, and second, that the museum has the world's most important scientific collection of craniums from domestic dogs and is carrying out research on the Swiss mountain dogs.[49]

Barry's biography as a museum object can be conveyed along two lines, one following his hide, a second following his skull. Together they lead to the scientific construction of the historic dog made in 2000 by curator and archaeozoologist Marc Nussbaumer, taxidermist Christoph Meier, and scientific illustrator Niklaus Heeb. Barry's skin had already been stuffed three times: immediately after the dog died in 1814, once more in 1826, and again in 1923. No picture survives of the first mount that was to be seen in the Museum der Naturgeschichte Helvetiens, the natural cabinet that predated the foundation of Naturhistorisches Museum der Burgergemeinde Bern in 1832. A new preparation was made in 1826 by technical assistant Hans Caspar Rohrdorf, showing Barry in a humble position that could be interpreted as representing the dog's devotion to humans. William Youatt considered Barry the highlight of the Bern collection, referring to the dog "as the noble quadruped whose remains constitute one of the most interesting specimens in the museum in Bern."[50] After one hundred years, however, *Hutchinson's Dog Encyclopedia*

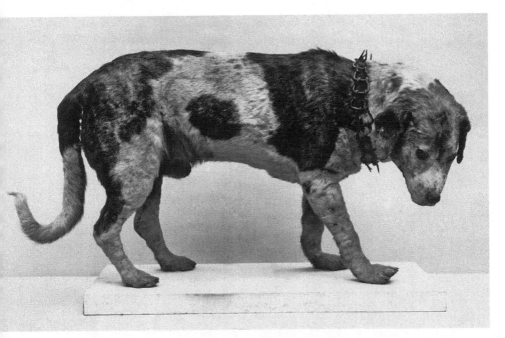

FIG. 6.2 Barry before new mounting, 1923. Photo © Natural History Museum of Bern.

"characterized him as a delight to the historic mind and a mild shock to the aesthetic eye." Not only was the skin poorly mounted, but it was evident that Barry bore little resemblance to the modern Saint Bernard breed. His skull was too long and without the breed's characteristic stop, his legs seemed weak, his coat texture was more wirehaired than rough and smooth, and the colors of his head were incorrect.[51]

By 1923, the stuffed Barry was badly deteriorated (fig. 6.2). The hide was cracking and the body was collapsing—"a travesty of how Barry was imagined."[52] Taxidermist Georg Ruprecht made a "new" Barry in 1923 and mounted the skin using the dermoplastic method in which the skin was arranged on a life-true mannequin (fig. 6.3). But Ruprecht had no pictures of Barry for a model. The only remains of the dog were the skin, skull, and perhaps some parts of the skeleton. Instead of using the bone material to settle the proportions of the historic dog, Ruprecht remodeled the dog according to the standards of the Saint Bernard breed that had been constructed after Barry's lifetime.[53]

Marc Nussbaumer has shown that the Saint Bernard dogs were far from homogeneous, and that the discussions on what the breed should look like focused on the shape of the skull. Nussbaumer thinks that it would have been possible to consider Barry, the most famous of the hospice's rescue dogs, as a

FIG. 6.3 Barry after Ruprecht's new mounting, 1923. Photo © Natural History Museum of Bern.

standard for the breed.[54] Instead, Ruprecht did the opposite. In the 1920s, a perfect Saint Bernard was to express benevolence, dignity, intelligence, strength, and endurance. The ideal head was large, massive, and the stop was abrupt and well-defined, the muzzle straight from nose to stop. The forelegs were to be of good length, perfectly straight and strong in bone, and the hind legs muscular.[55] The head was given most attention and the highest points in the ring. The Barry we see today reflects this aesthetic.

Conversely, Ruprecht's mount resembles a painting from about 1695 of a hospice dog with a clearly defined stop; the museum possesses an eighteenth-century copy of the painting. Nussbaumer argues that Ruprecht probably used the painting as a model for his work.[56] It may be possible that Ruprecht knew the drawing, but it was not in the museum's collection before 1933. Ruprecht neglected Barry's skull but paid homage to both the hospice

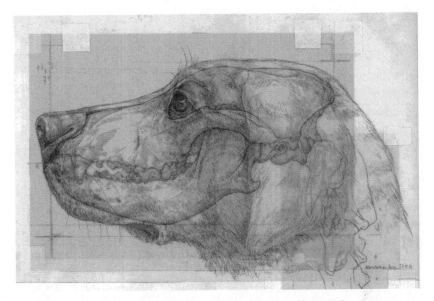

FIG. 6.4 Reconstruction of Barry's head, 2000. Drawing by Niklaus Heeb;
© Niklaus Heeb.

dogs and to the Saint Bernard standards that favored heavy, broad heads and
long legs.[57] Thus the new Barry anchored the breed standards to the history of
the famous rescue dogs, rather than the reverse. This Barry is still to be seen in
the museum's entrance hall. He holds his head in an upright, alert position, as
if he is looking for rough weather and snowslides, and his posture performs a
very different "dogness" compared to that of the old, humble Barry.

## Myth and Matter

When even natural history preparations don't represent the real thing, what
can we trust? What truths are depicted—or expected—from stuffed animals,
and where do these expectations originate? Is it possible to grasp the historic
Barry at all? Barry's bicentennial anniversary was celebrated by reconstructing
him once more—but this time only on paper. By using the skull as the basis
for a model of his head, scrutinizing the hide to find the right proportions,
studying the photograph of Barry before Ruprecht had mounted him, and
looking for differences in the texture of his coat, the historic Barry was reconstructed on paper. The research team discovered that not only had the head
been modeled with a stop, but the distance between the eyes had also been
broadened to make it more massive. The "Elbogenschwiele"—the wear of the

coat on the elbows—had been placed too high on the legs. This means that the mount from 1923 is about ten centimeters taller than Barry had been, making the dog nearly as tall as he is long, proportions that are absent in all dog breeds.[58]

The most authentic version of the physical Barry (in terms of his outward appearance) is the reconstruction made in 2000 (fig. 6.4), but even this stands as a questionable attempt to produce an accurate recreation of the real dog. Clever taxidermists know that even if the aim of their work is to reshape lived form as true to nature as possible, an animal mount will necessarily be an artifact and as such be invested with intention and meaning. According to Meier, the Barry we see in the entrance hall is not only an example of extraordinary taxidermy work; what Ruprecht created was a monument of an animal personality, one among few.[59] Barry's personality had been molded and remolded throughout the hundred years after the dog died to when the new mount was made. Perhaps this is the essence of the stuffed Barry: a memorial to the notion of the hero dog, and with him the nimbus of the Saint Bernard breed.

Now disassembled, the skull and skin belong to different sections of the museum. The first step in transforming an animal into a specimen in a natural history collection is to inscribe it in the museum's general accession register. The accession number lifts the item out of the everyday circulation of things, annexes it to the multitude of objects in the museum collections, and places it in new material and intellectual contexts.[60] Museum objects are collected because they are considered valuable: they may be unique, rare, outstanding, typical, or representative. Barry's skull is one of two thousand dog skulls belonging to the Albert Heim Foundation, one of many that are used to demonstrate the development of the Saint Bernard dog during the last two hundred years.[61]

Since the eighteenth century, descriptions of wild specimens have been based on an exemplar or series of examples. Scientific names are rooted to descriptions of particular specimens. During the following century, original type specimens were moved to museums' scientific collections.[62] A type specimen is the standard for a name, naming being based on literary practices of description and publication. The importance and fame attached to the discovery and description of a type specimen is discussed in chapter 5.[63]

Barry's history challenges this practice with regard to the Saint Bernard breed. If Ruprecht had decided to maintain old Barry's head and his short legs, considered against the 1923 breed standards, could Barry have served as a type specimen? Biologically, all dogs are varieties of the gray wolf (*Canis lupus*), domesticated over tens of thousands of years. But dog breeds were invented in the nineteenth century and are not defined by type specimens. Instead, dog breeds are defined by standards described and adopted by cynological associations and assemblies. Like standards in general, breed standards are not static but will be interpreted and redefined. This may explain why natural history museums rarely have domestic forms in their collections. When canine breeds and their standards were introduced in the second half of the nineteenth century, standards told less about the beast's natural form than about an aesthetic interpretation of dog, history, and myths. Rather than conserve a form, breed standards have invited interpretations and alterations of the animal's exterior, sometimes to a degree demonstrating the aesthetics of exaggerations.[64]

The museum in Bern has in many ways tried to punctuate the popular image of the Saint Bernard with a liquor barrel around its neck and to tell a counter-story, in which Barry is presented as an extraordinarily clever dog that always ranged the mountains in company with a monk or a mountain guide—without a barrel. Originally, the stuffed Barry carried a spiked collar to protect his neck and throat from the attacks of wild beasts. The Barry of 1923 received a barrel fastened to his collar. In 1978, the barrel was removed and sent to the museum in the Saint Bernard hospice. It had been made long after Barry died. Did the big working dogs in the Alps ever carry barrels on their collars? Probably not, because the barrel would have hampered the dogs while working.[65] But facts may yield to myths even in natural history museums, and in honor of his two hundredth anniversary, the barrel was once more put on Barry for a photograph.

In the entrance hall, lifted on a pedestal in his new glass case made in 2005, hermetically sealed and continuously controlled to prevent skin and coat deterioration, Barry is the first animal visitors meet in the Bern Natural History Museum. He is declared the museum's darling, and some visitors come only to see the dog. This mounted Barry belongs certainly more to the realm of imagination than to that of science. Stuffed animals in natural history museums are artifacts: they are, mostly, replaceable, three-dimensional nature illustrations, representing a species and not an individual animal. The transformation of Barry in 1923 materialized the imagined Saint Bernard and made it a tangible truth. Even if Barry's coat has been rearranged three times, the mounted dog retains an aura of authenticity. His aura is made of legends stemming from the historic dog paired with the fame of the breed. Barry epitomizes the loyal and faithful dog in service of man. For nearly two hundred years, the most famous Saint Bernard in the world resisted being reduced to a number in a museum collection. Today he is number 1050137 in the general register. Regardless of this, Barry still keeps his singularity as the relics of a historic dog, a representative of the world-famous Saint Bernard breed and a representation of the Saint Bernard's contribution to Swiss identity.

## NOTES

1. Daston, "Introduction," 20.
2. Meisner, "Barry," 25: "von nackten, unfruchtbaren, mit ewigen Schnee bedeckten Felsen."
3. Nussbaumer, *Barry vom Grossen St. Bernhard*, 39.
4. Marquis, *Saint Bernard*, 12.
5. Nussbaumer, *Barry vom Grossen St. Bernhard*, 32.
6. Ibid., 30–35.
7. Ibid., 40.
8. Ibid., 35.
9. Ibid., 36. *Hutchinson's Dog Encyclopedia*, 3:1509–10, maintains that the monks began with the Swiss Sennenhund, and that the hospice dogs were later interbred with the Great Dane, the Newfoundland, and the Pyrénean.
10. See http://en.wikipedia.org/wiki/Canine_terminology (accessed March 28, 2011).
11. Marquis, *Saint Bernard*, 45.
12. Meisner, "Barry," 26: "Zwölf Jahre lang war er unermüdet tätig und treu im Dienst der Verunglückten, und er allein hat in seinem Leben mehr denn 40 Menschen gerettet. Der Eifer, der er hierbei bewies, war ausserordentlich. Nie liess er sich and seinen Dienst mahnen. Sobald der Himmel sich bedeckte, Nebel sich einstellten oder die gefährlichen Schneegestöber sich von weitem ankündeten, so hielt ihn nichts mehr im Kloster zurück; nun strich er rastlos und bellend überall umher und ermüdete nicht, immer und immer wieder nach den gefährlichen Stellen zurückzukehren, ob er nicht irgendeinen Sinkenden halten oder einen Vergrabenen hervorscharren könne, und konnte er nicht helfen, so setzte er in ungeheuren Sprüngen nach dem Kloster hin und holte Hilfe herbei. Als das edle, treue Tier alt und kraftlos war, sandte es der würdige Prior des Klosters durch einen seiner Diener nach Bern, mit dem Wunsche, dass er nach seinem Tode, welcher im Jahre 1814 erfolgte, in unserm Museum aufgestellt werden möchte. 'Es ist,' schrieb der gefühlvolle Mann, 'mir angenehm und gleichsam ein Trost, zu denken: dass dieser treue Hund, der so vieler Menschen Leben gerettet, nach seinem Tode nicht so bald vergessen sein wird!'"
13. Thorsen, "Speaking to the Eye," 59–61.
14. On Meisner, see http://www.deutsche-biographie.de/sfz60068.html (accessed January 26, 2013). The oldest naturalia collections in Bern were kept in the city library and dated back to 1694. The natural history collections together with the botanical garden and the ethnographic collections were separated from the library in 1832 and became sections in a new museum. See Huber, "Die Geschichte des Naturhistorischen Museums Bern," 13, 19.
15. The popularization of natural history started in the eighteenth century by authors "who aimed to entertain as well as to instruct." See Thomas, *Man and the Natural World*, 282.
16. Irmscher, *Poetics of Natural History*, 8.

17. Römer and Schinz, *Naturgeschichte der in der Schweiz einheimlichen Säugethiere*, 20: "Nirgends, so viel wir wissen, werden sie in ein der so menschenfreundlichen Absicht gebraucht und gehört daher diese Thatsache ausschliessend in die Geschichte des schweizerischen Hundes."

18. Schinz, *Fauna Helvetica oder Verzeichniss aller bis jetzt in der Schweiz entdeckten Thiere*, 16: "Der letzte Abkömmling der ächten Stammrace rettete bei vierzig Menschen das Leben und ist auf dem Museum zu Bern aufgestellt."

19. See also Alberti, "Constructing Nature behind Glass."

20. Barbou, *Le chien*, 107.

21. Franklin, *La vie privée d'autrefois*, 93.

22. Youatt, *The Dog*, 13.

23. Ibid., 7.

24. Ibid., 38.

25. Hamerton, *Chapters on Animals*, 40, 41.

26. Clark, *Animals and Men*, 50.

27. Ibid., 45, 46.

28. Leonard, *The Dog in British Poetry*, x.

29. See http://en.wikipedia.org/wiki/Samuel_Rogers (accessed September 20, 2010).

30. Transcription following the epitaph on Boatswain's monument.

31. Probably the most famous painting demonstrating the emotional bond between the poor man's dog and its master is Sir Edwin Henry Landseer's work *The Old Shepherd's Chief Mourner* (1837). The anecdote of "Greyfriars' Bobby," a dog that spent the rest of its lifetime on its master's grave at Greyfriars' Kirkyard in Edinburgh, is a literal parallel to Landseer's painting.

32. Ash, *Dogs*, 2:593.

33. Barazetti, *Saint Bernard Book*, 4.

34. For more about Albert Heim, Swiss Alp dogs, and scientific research, see the Albert Heim Foundation for Canine Research in the Natural History Museum of Bern, http://www.nmbe.ch/research/vertebrates/research/kynologie/albert-heim-foundation (accessed January 26, 2013).

35. Heim, "Barry," 70.

36. Scheitlin, *Versuch einer vollständigen Thierseelenkunde*, 270: "Wäre ich ich ein unglücklich gewesen, du würdest mich nicht angemurrt haben."

37. Ibid., 269: "Du warst ein sinnvoller Menschenhund, mit einer warmen Seele für Unglüchliche."

38. Ibid., 270.

39. Ibid., 269: "Du bist das Gegenteil von einem Todtengräber. Du machst auferstehen."

40. Nussbaumer, *Barry vom Grossen St. Bernhard*, 50.

41. Ibid., 45.

42. Leonard, *The Dog in British Poetry*, 296.

43. Girardin, *Il cane nella storia e nella civiltà del mondo*, 104.

44. Nussbaumer, *Barry vom Grossen St. Bernhard*, 26.

45. Andreasen, *Sankt Bernhard Bogen*, 7. Andreasen builds on Barazetti, *Saint Bernard Book*.

46. Terhune, *Book of Famous Dogs*, 281–83.

47. Sanborn, *Educated Dogs of Today*, 17.

48. Thorsen, *Hund!*, 242.

49. Nussbaumer, *Barry vom Grossen St. Bernhard*.

50. Youatt, *The Dog*, 16.
51. *Hutchinson's Dog Encyclopedia*, 3:1513.
52. Nussbaumer, *Barry vom Grossen St. Bernhard*, 60.
53. The Swiss breed standards were accepted internationally in 1887. See ibid., 81. According to these, the "stop is clearly defined." Ash, *Dogs*, 2:600.
54. Nussbaumer, *Barry vom Grossen St. Bernhard*, 60.
55. Ash, *Dogs*, 2:611–12.
56. Nussbaumer, *Barry vom Grossen St. Bernhard*, 61.
57. Ibid.
58. Ibid., 63.
59. Ibid., 67.
60. Alberti, "Objects and the Museum."
61. About the morphological change of skull shape in the Saint Bernard breed, see Drake and Klingenberg, "Pace of Morphological Change."
62. About type specimens and their naming, see McGhie (chapter 5) in this volume.
63. See also Daston, "Type Specimens and Scientific Memory."
64. Thorsen, *Hund!*, 314–23.
65. Nussbaumer, *Barry vom Grossen St. Bernhard*, 53.

## BIBLIOGRAPHY

Alberti, Samuel J. M. M. "Constructing Nature Behind Glass." *Museum and Society* 6, no. 2 (July 2008): 73–95.
———. "Objects and the Museum." *Isis* 96 (2005): 559–71.
Andreasen, Ingrid. *Sankt Bernhard Bogen*. Denmark: Self-published, 1972.
Ash, Edward. *Dogs: Their History*. 2 vols. London: Ernest Benn, 1927.
Barazetti, W. F. *The Saint Bernard Book*. United Kingdom: Self-published, 1954.
Barbou, Alfred. *Le chien*. Paris, 1883.
Clark, Kenneth. *Animals and Men: Their Relationship as Reflected in Western Art from Prehistory to the Present Day*. London: Thames and Hudson, 1977.
Daston, Lorraine. "Introduction." In *Things That Talk: Object Lessons from Art and Science*, edited by Lorraine Daston, 9–27. New York: Zone Books, 2004.
———. "Type Specimens and Scientific Memory." *Critical Inquiry* 31, no. 1 (2004): 153–82.
Drake, Abby Grace, and Christian Peter Klingenberg. "The Pace of Morphological Change: Historical Transformation of Skull Shape in St. Bernard Dogs." *Proceedings of the Royal Society, Biological Science* 275, no. 1630 (2007): 71–76.
Franklin, Alfred. *La vie privée d'autrefois: Modes, moeurs, usages des parisiens: Les animaux*. Vol. 2. Paris: I. E. Plon, 1897.
Girardin, Carlo Umberto. *Il cane nella storia e nella civiltà del mondo*. Bergamo, 1930.
Hamerton, Philip Gilbert. *Chapters on Animals*. London: Seeley, Jackson, and Halliday, 1874.
Heim, Heinrich. "Barry: Tatsache und Dichtung." *Schweizerisches Hunde-Stammbuch* 20, no. 32 (1933): 65–76.

Huber, Walter. "Die Geschichte des Naturhistorischen Museums Bern." In *1832–1982 Festschrift zur 150-Jahr-Feier: Naturhistorisches Museum Bern Jahrbuch*, edited by Walter Huber, 11–49. Bern: Naturhistorisches Museum Bern, 1982.

*Hutchinson's Dog Encyclopedia*. 3 vols. London: Hutchinson & Co., 1935.

Irmscher, Christoph. *The Poetics of Natural History: From John Bartram to William James*. New Brunswick: Rutgers University Press, 1999.

Jesse, George J. *Researches into the History of the British Dog*. 2 vols. London: Robert Hardwicke, 1866.

Leonard, Robert Maynard. *The Dog in British Poetry*. London: David Nutt, 1893.

Marquis, Marcel. *Saint Bernard*. Genoa: Editions du Grand-Saint-Bernard, 1988.

Meisner, Friedrich. "Barry." *Alpenrosen* (1816): 25–26.

Nussbaumer, Marc. *Barry vom Grossen St. Bernhard*. Bern: Somowa Verlag, 2000.

Römer, Jakob, and Heinrich Rudolph Schinz. *Naturgeschicte der in der Schweiz einheimlichen Säugethiere*. Zurich: Heinrich Gessner, 1809.

Sanborn, Kate. *Educated Dogs of Today: An Illustrated Record of Canine Intelligence Marking an Advance with the Modern Movement of Man*. Boston, 1916.

Scheitlin, Peter. *Versuch einer vollständigen Thierseelenkunde*. 2 vols. Stuttgart and Tübingen: J. W. Cotta, 1840.

Schinz, H. R. *Fauna Helvetica oder Verzeichniss aller bis jetzt in der Schweiz entdeckten Thiere*. Neuchâtel: Petitpierre, 1837.

Terhune, Albert Payson. *A Book of Famous Dogs*. New York, 1937.

Thomas, Keith. *Man and the Natural World: Changing Attitudes in England, 1500–1800*. New York: Oxford University Press, 1996 [1983].

Thorsen, Liv Emma. *Hund! Fornuft og følelser*. Oslo: Pax Forlag, 2001.

———. "Speaking to the Eye: The Wild Boar from San Rossore." *Nordisk Museologi* 2 (2009): 55–80.

Wood, John J. *Man and Beast Here and Thereafter*. New York: S. Harper and Brothers, 1875.

———. *Stories and Anecdotes of Dogs*. Philadelphia: Willis P. Hazard, 1856.

Youatt, William. *The Dog*. Philadelphia, 1852.

# PART III

INTERACTING

Popular Entomology and Anthropomorphism in the Nineteenth
Century: L. M. Budgen's *Episodes of Insect Life*

*Adam Dodd*

"There are more ways than one," wrote Philip Henry Gosse in 1860, "of studying natural history."[1] For those of us interested in the genealogy of natural history, and in human-animal relations, this point is essential. Although known
for presenting a staggeringly eccentric view of Earth's history in his *Omphalos*
(1857), Gosse nevertheless displayed some good sense in his summary of the
ways that natural history might be studied—a summary that echoes current
perspectives on the historical diversity of the "doing" of natural history itself:

> There is Dr. Dryasdust's way; which consists of mere accuracy of defini-
> tion and differentiation; statistics as harsh and dry as the skins and bones
> in the museum where it is studied. There is the field-observer's way; the
> careful and conscientious accumulation and record of facts bearing on
> the life-history of the creatures. . . . And there is the poet's way; who
> looks at nature through a glass peculiarly his own; the aesthetic aspect,
> which deals, not with statistics, but with the emotions of the human
> mind,—surprise, wonder, terror, revulsion, admiration, love, desire, and
> so forth,—which are made energetic by the contemplation of the crea-
> tures around him.[2]

The contemporary field of animal studies has turned increasing attention to the ways in which naturalists have historically described animals
and animal behavior, provoking a range of questions about the role that

discursive conventions fulfill in the apprehension and conception of animal life. As Eileen Crist has argued, although many of the accounts provided by naturalists are to be considered unscientific in comparison to those of classical ethologists, this does not render all naturalist accounts redundant.[3] Although such accounts may be seen as subjective (or in a more antiquated parlance, "poetic"), they can serve as opportunities to investigate how notions of subjectivity and objectivity have historically formed around the observation and description of animals, rather than as cues to disregard the accounts altogether. Moreover, a close examination of the data offered by naturalists may (and often does) reveal useful and intimately detailed recordings of animal behavior systematically omitted by classical ethological methods, such as singular or unique instances, or moments principally significant in the aesthetic impact they impart to the observer. The complete absence of "subjective" and anecdotal accounts from compiled descriptions of animal behavior can ultimately lead to a view of animals as what Crist describes as "mechanomorphic objects," motivated only by biological imperatives, leading lives that are little more than predictable (albeit often complex) patterns of action that remain ultimately meaningless to the actors themselves. While such a perspective may effectively shield the human observer from the allures of anthropomorphism, it will inevitably lead to a deficient, condescending vision of the animal being observed and described.

Few animals have been as subject to this vision as have the insects. Indeed, in much entomological media produced since the early twentieth century, insects (especially social insects) are routinely presented as mechanomorphic objects driven purely by instinct, "hard-wired" to perform specific tasks— as more "robotic" than animal. On a broader cultural level, it is apparent that little, if any, value is attached to the lives of individual insects, partly because insects are themselves rarely considered as individuals. To consider an insect as an individual, as a sentient creature possessing unique behavioral traits, for example, is generally regarded as "excessively anthropomorphic," even while debates about what precisely constitutes excessive anthropomorphism remain unresolved.

In this chapter, I discuss an exemplary historical case of the anthropomorphic description of insects, L. M. Budgen's three-volume *Episodes of Insect Life* (1848–50),[4] in order to shed more light on the origins, dimensions, and effects of the anthropomorphic portrayal of insects. Many of today's commonplace anthropomorphic portrayals of insects have their origins in representational conventions that first gained popularity in the nineteenth century. As we will see in the discussion that follows, when performed capably, the episodic and anecdotal description of insect activity facilitates an anthropomorphic view of insects as the inhabitants of an "insect world" and largely eschews interpretations of them as mechanomorphic beings motivated completely by forces beyond their comprehension or control. With such descriptions, real events occurring in the natural world come to be seen as comprising actual narratives enacted by players (rather than as "discursive constructions") and are considered to be of genuinely moral, allegorical, and instructional character. For many nineteenth-century naturalists (as with a number of their early modern predecessors, such as Rösel, discussed by Brian Ogilvie in this volume), events occurring within the perceived insect world effectively "spoke back" of moral, allegorical, and instructional truths: insects were inherently emblematic. Hence, this chapter treats Budgen's work not merely as an example of "how people *assign* meaning *to* the world," but also of "how people *receive* meaning *from* their world."[5] More broadly, it offers Budgen's work as but one example of the various rhetorical techniques nineteenth-century writers employed in order to engage the public with knowledge about insects—and to sell their books in a rapidly expanding marketplace.

## An Introduction to *Episodes of Insect Life*

In the middle of the nineteenth century, an especially rich example of the episodic description of insect behavior, and its anthropomorphic effects, appeared in L. M. Budgen's *Episodes of Insect Life*. A three-volume work, with each volume describing four consecutive months of insect activity, it offers the reader an entrance to the insect world, and a whimsical journey through a year within it, that is as educational as it is fanciful.

Little is known of the book's author, who published under the pseudonym of "Acheta Domestica, M.E.S."—*Acheta domestica* being the Latin name of the European house cricket. Her other works include *March Winds and April Showers* (1854), *May Flowers* (1855), and the somewhat hallucinatory fireside companion *Live Coals; or, Faces from the Fire* (1867). Each of the three annual volumes of *Episodes* was consecutively released, just ahead of Christmas, available in both monochrome and hand-colored editions.[6] They all received overwhelmingly positive reviews from the critics. A collection of excerpts from contemporary reviews is to be found in the back matter of Alice Carey's *Clovernook; or, Recollections of Our Neighbourhood in the West* (1852), where it is reported that "no work published during the year, has received so extensive and favorable notices from the British Quarterlies and Newspapers as the Episodes of Insect Life." Not only are the excerpts reproduced here extremely favorable, but they also illuminate just what it was about this work that captured the imagination of readers in the United Kingdom and across the Atlantic in the middle of the nineteenth century. The press praised *Episodes* for its educational, instructional, absorbing, and entertaining portrayal of insect life that (re)enchanted the natural world for the benefit of adults and children alike. The *Boston Post* reported that "it is a beautiful specimen of book-making. The character of the contents may be already known to our readers from the long and very favourable attention they have received from the English reviewers. The illustrations are at once grotesque and significant." The *Rochester Daily Democrat* wrote, "The style is the farthest possible remove from pedantry and dullness, every page teems with delightful matter, and the whole is thoroughly furnished with grace and beauty, as well as truth. One giving himself over to its fascinating charms, might readily believe himself fast on to the borders, if not in the very midst of fairy land." The *Ontario Repository* called it "wonderfully beautiful, graceful, and entertaining. Children can read it with understanding and be enraptured by it; and this is no small thing to say of a work not especially intended for juveniles." The *Christian Intelligencer* observed that "we have in this work deep philosophy and an endless flow of humor, and lessons set before us, drawn from ants, beetles, and butterflies, which we might do well to ponder." Back in the United Kingdom, the *Morning Chronicle* had remarked that "the whole pile of Natural History—fable, poetry, theory, and fact—is stuck over with quaint apothegms and shrewd maxims deduced, for the benefit of man, from the contemplation of such tiny monitors as gnats and moths. Altogether, the book is curious and interesting, quaint and clever, genial and well-informed"—while the *Sun* had described how "never have entomological lessons been given a happier strain. Young and old, wise and simple, grave and gay, can not turn over its pages without deriving pleasure and information." *Episodes of Insect Life* was, in the opinion of the press, an ideal insect book with which to amuse and instruct all readers, motivating a flow of "fairy-tale natural history" from Europe to North America in a style that would become most readily associated with Disney productions in the twentieth century. The series was published in the United States in 1851, and an abridged, single-volume version, edited and revised by Rev. John Wood, was published in 1867, though it seems to have been met with less acclaim than were its predecessors.

The critical responses to *Episodes* point to an important quality that distinguishes Budgen's work from established genres of the period—its novel ability to amalgamate an adult and juvenile readership at a time when the distinctions between adults and children were themselves becoming increasingly pronounced. The subject of insects was itself ideally suited to this transgenerational appeal, but Budgen wove elements of two very popular contemporary genres—fairy stories and animal stories—into her writing. While the insect world and its inhabitants were presented as a type of nostalgic fairyland in the nineteenth-century style, the author's "alter ego," Acheta Domestica, exhibited the classic features of an anthropomorphized character from contemporary animal stories. As Cosslett observes, "Like the fairy story, the animal story has migrated down the hierarchy of literary genres from adults to children, in consequence of an increasing polarization between adults and children. Adults were more and more seen as rational and cultured, while children were imaginary and primitive. At the same time, Enlightenment educational thought differentiated between fairy stories and animal stories, the one encouraging fear and suspicion, the other encouraging benevolence and knowledge."[7]

*Episodes of Insect Life* does not conform to such distinctions, and perhaps this explains the success it enjoyed at the time. In line with contemporary animal stories, which were not "just a repetition of the Aesopian fable, but added either an anti-cruelty message and/or natural historical information to the fabular genre,"[8] *Episodes* encouraged a mindful and compassionate view of animal (and specifically insect) life. But like fairy tales, it also exhibited an evocative longing for something lost, an idealized, vaguely medieval past—a type of collective infancy—when the human connection to nature was unbroken by the ascendency of mechanized technology and the modern, industrialized city. In this sense, the insects of *Episodes* were employed to bridge the generic animal story and fairy tale, and thus to coalesce an adult and juvenile readership in an appreciation of the fairylike qualities of certain parts of the animal, or natural, world.

## The Narrative Images of *Episodes of Insect Life*

Each volume of *Episodes* displays the same gilded image on its cover. An anthropomorphized cricket (Acheta Domestica), dressed in the traditional garb of an early nineteenth-century gentleman, sits on a tree stump, an open book resting on one leg, with two more volumes on the grass in front of him. These, we are to assume, represent the three volumes of *Episodes* itself. Acheta's right arm is raised in a gesture of instruction. Framing Acheta in the center of the image is an assortment of insects, including beetles, a cricket, butterflies, and a bee. At the top of the frame, a spider rests in a web, positioned above the intimate setting like the sun, or perhaps the moon. The insects seem curiously to approach Acheta as his lesson begins, and draw the reader closer to the book itself. The quotation at the bottom reads, "He filled their listening ears with wondrous things" (see fig. 7.1).

This narrative image effectively portrays a kind of insulated, secret space within an insect world where these "wondrous things" may be imparted, and where the subsequent enchantment may transpire. The curiosity of the general reader, not yet having been introduced to the central character and author (who, in hindsight, bears some similarity in both form and function to Disney's Jiminy Cricket), is peculiarly aroused by this grotesque yet strangely inviting scene, which, despite a latent monstrosity, conveys a palpable benevolence. What secures this benevolence is the foregrounding of the impartment of knowledge as the pretext for the congregation of radically different, and potentially adversarial, life forms; the scene is one of a peaceful lesson that seems to transcend (if only for the duration of the lesson) the natural law of "eat or be eaten." The relationship of nature and the book, and of the infant's "story time," is significant here, for the book is what makes this scene of natural communion possible; it is the foundation of this reconfiguration of nature into a classroom. Acheta seems to be addressing both the adult and juvenile reader, and the insects themselves, and in this way human and insect become an amalgamated audience, a union of species embodied by Acheta at the center of the frame. The reader is encouraged to disregard, or at least to assuage, the fixity of their human subjectivity. Set at the scale of the insects, the perspectival position of the human observer is one that would appear in many subsequent texts that sought also to convey a sense of human immersion in "the insect world."

Like the cover image, the title vignette—"So issue forth the seasons"— is repeated across the three volumes. Illustrated with four anthropomorphized insects forming a traditional roundel, and dressed accordingly, the author describes it thus: "First we have WINTER in his merriest mood, represented by the Cricket, bedecked with Christmas holly, and alive with fun and jollity. By his right hand he holds the Brimstone Butterfly, emblem of SPRING, primrose of papilions in habits and in hue. Beneath, the jocund Grasshopper linked to the above by a vernal wreath, figures the bright SUMMER and in the glowing Peacock Butterfly, rich in her velvet train as the autumnal flowers she frequents, we welcome AUTUMN, bearing the ripe sheaf and presenting her merry associate with the fruit of the vine."[9]

FIG. 7.1 Secret lessons in nature: the alluring gilded cover to L. M. Budgen's *Episodes of Insect Life*.

This image (in which the four seasons of the year are *arthro*pomorphized) portrays Budgen's playful chronicling of one complete year of insect life across the three volumes: volume 1 describes January through April and contains chapters 1–17; volume 2, May through August across chapters 18–34; and volume 3, September throughDecember, containing chapters 35–51. Like the cover image, the title vignette signals an entrance to an insect world in which violence, parasitism, and death are virtually absent—replaced by a sublime vision of a benevolent, transformative nature in which insects are seen as "little people" (see fig. 7.2).

In volume 1, the title vignette sits opposite a frontispiece portraying "butterflies in general," described by the author as "various species" that, "just risen and bursting from their chrysalidan shrouds, mount towards the skies or repose upon everlasting flowers" (1:xi) (see fig. 7.3). In contrast to the fanciful title vignette, this image is entomologically accurate, and a fairly extensive description of each butterfly, including its Latin nomenclature, is provided in the table of contents. On the surface, it is a particularly straightforward, objective representation of miscellaneous butterflies that portrays their various stages of metamorphosis. But the difficulty faced by many nineteenth-century entomologists in considering insect metamorphosis as a resolutely mundane, biological phenomenon is also signified here. Captioned with a passage credited to Longfellow, the symbolic relevance of metamorphosis is acknowledged, along with the juvenile appeal of the subjects: "And with child-like credulous affection / We behold those tender wings expand / Emblems of our own great resurrection / Emblems of the bright and better land."

The angelic connotation of insects' wings (by this stage beginning to merge with those of the emergent miniature fairy), their representation of the ability to move between worlds or states, is significant. Indeed, so prevalent was the spiritual significance of insect metamorphosis for the Victorians that Budgen, after the entomological explanation of the image, writes that "to the symbolic meaning of this picture it is scarcely needful to point, for in the Book of Nature, so truly described to be a Book of Emblems, the history of the fugacious Butterfly, as typifying the flight of the immortal soul, stands foremost for clearness, for exactitude, for beauty, and for solemn import" (1:xi–xii). Such a passage indicates the unfeasibility of an evenhanded assessment of all species of insects within Victorian entomology. Not only were some insects more beautiful than others, but some—particularly butterflies—were deemed emblematically and spiritually superior. Butterflies had been linked through Greek mythology to the immortal "psyche" (or soul, spirit, or mind) and, at least

"So issue forth the Seasons"_Spencer.

FIG. 7.2 Title page to all three volumes of Budgen's *Episodes of Insect Life*, in which the seasons are celebrated and personified by insects.

since Apuleius's *Metamorphoses* (or *The Golden Ass*), to the feminine—a motif exploited by many nineteenth-century painters such as Waterhouse and Bouguereau. The metamorphosis of the butterfly, and particularly its visible attainment of a "perfect" or "final" state—the imago—so far removed from the larval, mummified, or death-like state preceding it, aptly embodied the metaphysical precept (and mythology) of spiritual evolution. We see this notion illustrated later in volume 1, in a scene showing Acheta in the "catacombs" with a lantern, inspecting the butterflies in their larval stage.

And with child-like credulous affection,
We behold those tender wings expand,
Emblems of our own great resurrection,
Emblems of the bright and better land'.— *Longfellow*

FIG. 7.3 Frontispiece to first volume of Budgen's *Episodes of Insect Life*, showing butterflies in various stages of metamorphosis.

"So issue forth the Seasons"—*Spencer.*

FIG. 7.2 Title page to all three volumes of Budgen's *Episodes of Insect Life*, in which the seasons are celebrated and personified by insects.

since Apuleius's *Metamorphoses* (or *The Golden Ass*), to the feminine—a motif exploited by many nineteenth-century painters such as Waterhouse and Bouguereau. The metamorphosis of the butterfly, and particularly its visible attainment of a "perfect" or "final" state—the imago—so far removed from the larval, mummified, or death-like state preceding it, aptly embodied the metaphysical precept (and mythology) of spiritual evolution. We see this notion illustrated later in volume I, in a scene showing Acheta in the "catacombs" with a lantern, inspecting the butterflies in their larval stage.

FIG. 7.3 Frontispiece to first volume of Budgen's *Episodes of Insect Life*, showing butterflies in various stages of metamorphosis.

## Advocating the Mindful Appreciation of Insect Life

Apart from the reiteration of rather solemn spiritual themes and motifs drawn directly from nature, *Episodes* sought to amuse the general reader as a way of provoking interest in entomology as science. Dedicated to William Kirby and William Spence (authors of the widely esteemed *Introduction to Entomology*, published between 1815 and 1826), and to the nineteenth-century British naturalist and caricaturist Edward Forbes, the book's position between entomology and entertainment is articulated in the preface:

> The following Essays have been written, not with a view of teaching Entomology as a science, but of affording such a measure of acquaintance with the habits of the Insect world, as may serve to promote the ulterior and more useful design of cultivating the rudimental seeds of systematic investigation. For this, with many, sufficient leisure, fitting residence, and other appliances may be wanting, but few can entirely lack opportunity for becoming more observant of Nature's wonders, more impressible to her influences and her teachings, or more alive to the superior intelligence visible in her works. On nothing, perhaps, are the signs of that intelligence more obviously impressed than on the operations of Insects, which, as creatures pre-eminently under the rule of instinct, attest as pre-eminently that "The *mind* which *guides* them is divine." (1:vii)

To some degree, Budgen managed to reconcile a pantheistic idea of an omnipresent divine mind in nature with the basic tenets of natural theology in a Judeo-Christian context. Although studying nature as a way to know and edify God's works was not an uncommon pursuit in the nineteenth century, in the case of Budgen (and other proponents of "spiritualized" entomology, such as Rev. John Wood) we have the rhetorical description of insect activity as comprising narrative episodes, placed before human observers, by the Creator, for human benefit. This mode of description, it can be said without risk of overstatement, opens the floodgates for a range of powerful and often persuasive anthropomorphic representations. Within the preface alone, Budgen encompasses virtually all the fundamental traits typical of nineteenth-century anthropomorphic portrayals of insects. From her teleological, pantheistic position, a deeper significance of insect behavior becomes apparent; what is required is a shift in consciousness that in turn alters one's perception of insect life:

Thus contemplated, the constructive skill, selecting judgment, and seeming foresight of these tiny agents, as applied to the preservation of themselves or offspring, are exalted into themes of surpassing interest; and, as in all created things there exists a purpose out of and above themselves, it is evident in these displays of instinct, that the same informing principle which serves in its operation to direct the animal actor, is intended by its exhibition to amuse and to instruct the rational spectator. . . . It may perhaps become apparent that allegoric fable, poetic association, and moral analogy, are no forced productions, but only the luxuriant growths (leaf, flower, and fruit) of that branch of the tree of knowledge which belongs to Insect history. (1:vii–ix)

Again, Budgen emphasizes the inherent purpose and "informing principle" that directs the "animal actor," but here specifically interprets the purpose informing their display as something intended "to amuse and to instruct the rational spectator." This is a suggestion that insects exist in no small part for the benefit of human spectatorship (a "purpose out of and above themselves"), and so interpreting insect activity in terms of allegorical fable, poetic association, moral analogy, and aesthetic appreciation is not strictly a human imposition onto nature, but rather an openness to the natural growth of the "tree of knowledge which belongs to Insect history"—not simply an assignment of meaning *to* the world, but more a reception of meaning *from* the world. Budgen makes clear in the first pages of *Episodes* that none of the observations that follow will operate wholly outside of metaphor and that, in fact, metaphor is a natural and desirable mode of perception to adopt when seeking to describe and understand insect activity. This notion is solidified by insect history itself being considered metaphorically—that is, as a branch of the "tree of knowledge."

## The Cricket as Author

In the first chapter of volume 1 of *Episodes*, "The Cricket: Introductory," Budgen comments on the (relative) abundance of books on entomology already in print, such as those by Kirby and Spence, James Rennie, William Jardine, Hermann Burmeister, and John Obadiah Westwood. But she notes that "seeing how generally even these are left to tarnish on the shelf, something would seem to be required as an incentive to their more frequent handling" (1:4). This position runs throughout the book, with Budgen reminding the reader that its purpose is merely to guide them toward the published authoritative texts, and to entomology as a systematic practice. *Episodes* was a work that Budgen had long considered writing, but had delayed for uncertainty about how it should be structured. She also faced the problem, shared by many early popularizers of entomology in the nineteenth century, of widespread disregard for insects themselves, writing that "the first anxiety of a writer is, as all the world knows, to establish a kindly sympathy between himself and his readers; but how can this be speedily created betwixt one who, as an Entomologist, would seem to think of nothing but Insects, and 'the many' who have always regarded them as below a passing thought?" (1:5).

Budgen was aware that entomological knowledge was indispensable for a deep and lasting appreciation of insect life, and she was also responding to an emergent audience of recreational readers interested in natural history. These were not only the learned adult readers whose copies of Kirby and Spence, Rennie, and Jardine sat tarnished on the shelf, but also juvenile readers who became, from around 1850, recognized as a readership in their own right. For Budgen, a collection of analogies seemed an effective way of inciting enthusiasm for insect life in readers of all ages, but the question of structure remained. Writing in 1847, there was no conventional format of popular books about insects for her to adopt. "Letters—Sketches—Conversations," she wrote, "these were familiar shapes into which our materials might be moulded; but they seemed, in one sense, too familiar; the public taste might be tired of these hacknied modes of dressing up the sister sciences." Budgen decided that "episodes might better serve our purpose, and impose fewer shackles on our roving fancy: Episodes, then, they shall be called—Episodes of Insect Life, providing every month a seasonable admixture of the Real and the Ideal" (1:6). Much like one of her subsequent books, *Live Coals; or, Faces from the Fire* (in which Acheta Domestica reappears), Budgen's chosen format of episodic narratives and analogies, interspersed with fine and fanciful illustrations, was ideally suited to leisurely perusal by the hearth, allowing readers to take in as much or as little as they pleased, with regular intervals provided for pause and reflection.

Budgen's decision to write episodically comes at the end of the year, and she then wonders how she will begin her writing of insects in the middle of winter, when "of all the summer myriads, the bulk have long ago expired; the remnant, scared even by the shadow of advancing winter, betook themselves to hidden places; and now old Christmas has benumbed them with his icy paw, and keeps them unconscious prisoners within the earth or waters" (1:6). The thought of how to go about this task leads to drowsiness, and Budgen falls asleep by the fire. She is awoken by the bells of the neighboring parish church, heralding the new year, and then left in the silence of her parlor once the peeling stops, "a silence which seemed deeper than usual, and more solemn, yet not to the spirit's ear unbroken" (1:7). Mingled thoughts, "of retrospect rather than prospect," rush through her mind, and overwhelm her idea for the book. Out of this overwhelming silence, however, comes the catalyst for her writing, and it is perhaps unsurprising that it takes the form of an insect, and specifically the insect's voice: "Of a sudden, however, it [the book] was again brought to the surface: a shrill sound broke upon the stillness; another chorus, within the house, succeeded to the hushed peal without. The Crickets, from the kitchen below, were uplifting their chirping strains to salute, in full concert, the new-come year. We were at no loss, now, for at least one cheerful subject wherewith to commence our Episodes.—Bless their merry voices for the opportune suggestion!" (1:8).

Budgen then takes up, not the pen, but the candle, and goes downstairs to find the crickets. (These are illustrated at the beginning of the chapter, with the clock in the background displaying the time of 12:13.) Most of them scamper away, but she captures "a straggler in the very act of draining the milk-pot, and [carries] him off to [her] parlour fire-side for the cultivation of a more intimate acquaintance, and with a view to making him as well-known to [her] readers, by sight, as he, or rather his merry fraternity are likely to be already by sound" (1:8).

At this point we are presented with an exceptional example of the transformation of the surrounding environment effected by the miniature, of the kind Susan Stewart has described in *On Longing*. Budgen places the cricket under a tumbler with a few crumbs of bread, and muses on the variety of foods enjoyed by its kind: "True . . . thou art not particular, 'scummings of pots, sweepings, bread, yeast, flesh and fat of broth,' thy pickings most esteemed, seem not, some of them, the most inviting fare; yet do not these dainties, each in its kind, serve to symbolize, not unaptly, the very sort of viands we would seek and set before our readers" (1:9). The things eaten by the insect, or that which is associated with it—its context—is becoming symbolic, and thus remarkable. Budgen becomes engaged with the significance of not just the cricket, but its surroundings, and a transformation begins to take place:

> For "scummings of pots," supposed we say the "cream of our subject," the most light, and withal, the richest of the agreeable matter already laid up by others, to be extracted by ourselves in the field of observation.

For "sweepings" let us put "gleanings,"—Gleanings of Entomology—
and we have the very term adopted by a well-known writer for his amus-
ing anecdotes in various branches of Natural History. Then "bread,"
with Cricket as with man, the very "staff of life," if poverty forbid him
not to grasp it, what substance more properly symbolic of that which
must form the ground-work of our book,—matters of solid fact, mixed
with and lightened by the "yeast" of illustration, discursive and pictorial. As for the "flesh" and "fat," the strongest fare on which the Cricket
delighteth to regale, may they not serve to typify that principle of mental nourishment, of all the most vital, afforded by the religious contemplation of all natural objects endowed with life. (1:9–10)

As Budgen approaches her first insect subject, her interpretation through
allegory and metaphor continues to shape the insect, its surroundings, and
ultimately its relation to herself. In its tendency to gnaw holes in wet woolen
stockings or flannel in the absence of water, Budgen sees the Cricket as "our
representative, as, thirsting after knowledge of our subject, we strive to extract
from it, even when seemingly most arid, a something of refreshing moisture"
(1:10). And finally, in the closing passages of the opening chapter's anthropo-
morphosis, Budgen observes that

> in all his doings, our Cricket is, confessedly, a pilferer, and taking, as we
> largely must, from stores collected by the labours and observations of
> others, we shall herein, also, resemble our prototype, except that we
> rob in open daylight, and thankfully acknowledge what we appropri-
> ate. There are yet other points of resemblance, more personal, between
> ourselves and the Cricket. As with him, a warm hearth in winter and a
> sunny bank in summer are the seats of our supreme felicity. Like him,
> also, we joy in the possession of a quiet retreat, and prefer to uplift our
> voice from behind a screen.
>
> We have now set forth quite as much of our design, and revealed as
> much of our personality as have come connected with our immediate
> subject, and from the scattered grains of intimation already dropt, some
> prying reader may even now have gleaned more about the Cricket's ways
> and whereabouts than we have thought it expedient to reveal. (1:10–11)

Budgen closes the chapter with an illustration of the amalgamation fabricated
by the preceding prose: herself, now transformed into an anthropomorphized

FIG. 7.4 Acheta asleep at the desk, after having devised *Episodes of Insect Life*.

cricket, asleep at the candle-lit desk, the first page of *Episodes* beneath her hand (see fig. 7.4). Budgen has transformed her authorial subjectivity in such a way as to function as an intermediary between worlds—the insect world and the human world. Her authorial form is deeply metaphorical, in line with her interpretation of insect life. Her symbolic assimilation of the cricket into herself represents the reader's (ideally) newfound sympathy for the cricket and for the insect subjects that follow in the succeeding chapters and volumes.

### The Points of Our Hobby: The May Fly as Symbolic of Entomology

The opening illustration to the second chapter of the first volume of *Episodes* continues Budgen's immersion of the reader in a metaphorical, or symbolic, interpretation of the insect world, functioning to symbolize entomology itself. It portrays the author, in anthropomorphized (though here more insectoid) form, attempting to catch a tiny, nondescript insect in a net while straddling the back of an enlarged May Fly as it flies over a lake. Budgen is attempting to represent entomology as a lighthearted pursuit that is aptly symbolized by the behavior of the May Fly: "Suffice it, now, that as in the Cricket we have introduced thee to our symbolic self, so in the May Fly we would beg thee to recognize our symbolic hobby" (1:12–13). Here, Budgen sets the tone for the investigations that follow, which seek to distance themselves from the cold, detached, and often morbid practices occurring in contemporary entomology. Moreover, as Nicola Bown has observed in her succinct examination of *Episodes*,[10] it is here that Budgen's fabrication of the insect world begins to coalesce most visibly with contemporary constructions of fairyland, and thus to constitute not just an imaginative escape from the increasingly industrialized city, but a distortion of the violence and death that typifies the everyday lives of insects.

Budgen attempts to reconcile what Bown has described as the "knowing" and "loving" gazes that functioned in many nineteenth-century observations of insects. Overall, however, Budgen is aligned with the "loving gaze." We can see this in her lengthy encomium of entomology, in which she emphasizes its capacity to shape the observer's experience of nature: "Dear Entomology! We have called thee our hobby, we have likened thee to a hack; but thou art more. Thou art a powerful Genie, a light-winged Fairy, not merely bearing us through earth, and sky, and water, but peopling every scene in every element with new and living forms, before invisible" (1:14).

The newly appreciated abundance of life at the subvisible level made the collection of insects appealing, and a well-filled cabinet edified (in a kind of microcosmic way) the grandeur of the Creation (see Ruud, this volume). But the cabinet and the collection also signified an enduringly contentious issue for entomology—the apparent cruelty inflicted on insect "specimens" in the name of the collection itself. As cabinet collections became increasingly fashionable and accessible, and taxonomies became more thorough, Nature's collectibles seemed virtually infinite, especially when it came to the most popular insect specimens—beetles, butterflies, and moths. Maintaining her position as mediator, Budgen comments on this aspect of her pursuit:

> Using our hobby as a hunter, we may pursue our game for two different purposes; that of scrutinizing living instincts, or arranging and looking at dead objects. . . . As for him whose delight in natural objects, of what kind soever, consists solely in their amassment, or is circumscribed within the walls of his cabinet, he is no naturalist at all, a mere kindred spirit of the Bibliomaniac, and little better than the miser whose iron heart is in his iron chest. Neither are specimens necessary to the study of Insects,

though, like the Hortus Siccus [dried, preserved garden] of the botanist, they are of great assistance, especially at its commencement. (1:23)

Budgen did appreciate, then, that the collection and cataloging of insects was not entirely purposeless, and that it could indeed assist in the advancement of knowledge about insects. But collected specimens were dead objects, and Budgen's primary incentive was to inspire enthusiasm for insect *life*. The question then becomes under what circumstances, and in what manner, should one take the life of an insect. Budgen writes that

> to take the little life even of a Butterfly is confessedly, and ought to be, matter of pain, and is, so far, a set-off against the pleasures of the Aurelian. Nor is it a set-off which use diminishes, for the more we notice the beauty of Insects and the more we learn of their movements, the greater becomes our reluctance to mar the former or arrest the latter by an unwilling hastening of the hand of death. It is only our moral right to do so on sufficient occasion for which we would contend. . . . On no principle can it be allowable to toy with torture. To take life quickly, and with far less suffering to the individual than what in the common course of nature it will for ever be liable to undergo, all must admit to be a different matter. (1:25)

After this definition of a justifiable method of execution (i.e., a relatively quick and painless death), Budgen then qualifies the killing of insects for pleasure by establishing the differing degrees of pleasure in Man and Insect. She does this by comparing "the sensual pleasures of a Butterfly versus the mental pleasure of a Man, such as can scarcely fail to be excited by a close examination of nature's miniature masterpieces of painting and mechanism" (1:25–26). Since there really is no comparison—Man's mental pleasure far exceeding anything within the possible experience of a Butterfly—the killing is justified. Still, the overall impression given by Budgen's treatment of this aspect is one of an enduring uncertainty, typical of popular nineteenth-century entomology's inability to figure insects neatly as either subjects or objects.

## "The People of the Hive": "The Royal Reform,—Bees as a Body Politic"

Having established the context for both her authorship and the pursuit of entomology, Budgen then proceeds on a seasonal, episodic journey through various encounters with insect life. These span fifty-one chapters, and space does not allow for discussion of them all. Some, however, are more noteworthy than others for their particularly anthropomorphic portrayals that highlight the ways in which the Victorian public was informed about insects, and the dilemmas faced by contemporary writers in representing insects and their activity.

A fundamental problem was educing methods with which to communicate the extremely unfamiliar physiology, habitats, and behavior of insects to readers almost wholly ignorant of these aspects of insect life. In many popular entomology books of the period, authors frequently included a definition of "the insect" itself within their introductions, and indeed this is still the case in many popular insect books today. For authors such as Budgen, whose spiritual or religious inclinations motivated the encouragement within her readers of deep compassion for insects, an effective tactic was to amplify obvious or superficial similarities between insect life and human life. This aspect of Budgen's work appealed to adults and children alike, since for much of the nineteenth century, adults and children were equally ignorant of entomology and thus could be addressed in much the same way by authors on the subject. The employment of anthropomorphism to educate children about animals, nature, and society has become such a common feature of children's literature since the final decades of the eighteenth century that it now seems quite naturalized. Considered in their historical and cultural context, however, Budgen's portrayals offer compelling insights into which particular aspects of nineteenth-century society and culture were seen as similar or harmonious with their insect counterparts, and with the natural world in general.

Of particular interest to the Victorians was the order Hymenoptera, which includes bees, ants, and wasps. It is perhaps unsurprising that the obvious similarities between human society and the most socially organized order of insects should be a source of interest for both nineteenth-century entomologists and the general public, and recent work has uncovered the depth of this interest.[11] Moreover, as Keith Thomas has noted, these parallels had been drawn with particular verve as early as the Stuart period, especially in attitudes toward bees.[12] Bees were employed to naturalize the prominent social hierarchies and class divisions of English society and reiterate the virtue of an unerring dedication to industry. Budgen provides an illustration of this trend in a two-part chapter of volume I titled "The Royal Reform,—Bees as a Body Politic," which traces the processes involved in the replacement of a queen within the hive.

Heading the chapter is the quotation "Subjects commonly do find, New made Sovereigns most kind," and it begins with a description of the death of a queen whose insect status, significantly, is not made immediately clear:

> There was great grief in one of the monarchies of the earth: the queen regnant of a numerous people had just been summoned to her ancestors. Yesterday she was a brilliant spark of life, from which light and activity extended to the very circumference of her kingdom; to-day, she is but a dull lump of mortality, casting its shade, and imparting its torpor far and wide around. The cheerful hum of labour is hushed in every quarter, and in its stead arises the mournful wail of lamentation. The royal corpse is cold, yet faithful attendants and devoted body-guards still watch around it, as if reluctant to believe their "occupation gone." Some of these loving creatures will even starve upon their grief, and fall dead themselves around the body of their defunct mistress. (1:232–33)

This passage shows the peculiar similarities that could be drawn between the beehive and European society, and the subsequent interpretation of bee behavior as indicative of emotional states, in this case, death as the result of grief. The bees are seen as "loyal" to the queen. The theme of royalty pervades the succeeding prose: we are told that "the kingdom of Apia (that of which we now write) was always a monarchy of marvels and of strange customs, and those which regarded the succession to the crown were some of the strangest among them" (1:233). Here Budgen begins to acknowledge the differences between the queen-making of the bees and that of human beings. Yet these differences simply magnify the sensation of looking into another world, a "monarchy of marvels and of strange customs" that is strongly reminiscent of fairyland.

Like fairyland, the monarchy of Apia is not entirely benevolent. As one older female bee explains to two male workers, frustrated with the simple replication of a monarch with the qualities of the last: "*We*, with our boasted elixir of certain and invariable properties, stuff and stimulate body and mind into an invariable shape, converting what would have been a useful, active member of society into an enormous, bloated, idle, cruel tyrant. *They*, by means of a wondrous art called mesmerism, acting on mind according as they please, contrive to expand the virtues and repress the vicious propensities of their infant subject (be it of royal or of humble birth), till they turn out of their moral laboratories paragons of princes and princesses, such as were never known since the world began" (1:242). Here, Budgen draws a bleaker comparison between humans and bees, including the presence and action of mesmerism, or "animal magnetism," in the maintenance of the social power structure. Indeed, Budgen's vision of the beehive is colored by the manipulation, murder, and repression inevitable in a complex modern society of multiple, conflicting interests.

In the next section of the chapter, however, titled "Bees as a Body Politic," Budgen attempts a more pleasing explanation of the bees' society:

> If any form of government be faultless, it must be one acting immediately under divine guidance, and of this class are the instinctive institutions of social animals, which are therefore perfect in their kind. Under an idea of such perfection (erroneously applied) the people of the hive have been held up to us people of the earth, not only as patterns of industry, but also of political economy, and have been cited not only as arguments for monarchy, but as models also of monarchical government. That men might, nevertheless, just as well attempt to build their cities after the pattern of a honey-comb, as to mould their institutions after those of the honey-comb's inhabitants, is evidence, we should think, in our little romance. (1:250–51)

Budgen is ultimately critical of the habit of holding up bees as models of political economy and government. The annihilation of three to four hundred drones from an average hive in the months preceding winter, when their jobs are done, warrants no emulation in human society: "Have those by whom Bee economy has been held up for human imitation, ever thought about the awful consequences which would be involved in even a partial copy of the above severely wholesome policy?" (1:255). Despite her apparently excessive anthropomorphism, Budgen has not completely forestalled recognition of the violence inherent in "bee economy." While the effectiveness of this economy cannot be disputed, Budgen's vision of humanity does not regard consummate political economy as the objective of human life or society. While recognizing the many behavioral similarities between humans and bees, it is the depth of the bees' engagement with particular behaviors that establishes their difference from us. And this depth of engagement signifies the presence, and importance, of instinct in shaping behavior itself, and in superseding (or precluding) morality. She closes the chapter with the following observation:

We may look again into the hive, but those who wish to dive deeply into the ways and wonders, the proceedings and policies of its busy inmates, must consult the works of Bee historians. Delightful pages some of them have written, reading much like human history, only more agreeably, because undefiled by moral blots. They tell us, it is true, that Bees go to war like human communities; that strong Bees rob the weak, like human villains; that angry Bees fight single combats, like human duel-lists; that Bees, well-fed and vigorous, will kill the old and helpless of their labourers. These are points of character, rough and sharp enough it must be owned; but they need not prick us in the reading, when we remember that Bees are but the passive elements of an unerring instinct. (1:263)

## Conclusion

This essay has examined only a few excerpts of Budgen's writing in the hope that these will sufficiently illuminate how the text as a whole is operating. But how are we to read it, in the twenty-first century, when nature has become understood as both a "construction" and as something we must save from ourselves? I have focused on how a nineteenth-century author conceived of human-animal authorship to construct and draw from a perceived "insect world," a lifeworld that exists at once in the natural world itself and within the imagination of the human observer. Much of what Budgen describes in her book invokes a view of insect life that operates outside the spectrum of objectivity as generally understood, yet in doing so, it does not abandon the pursuit of truth. Located in its historio-cultural context, *Episodes of Insect Life* stands as a lively work of popular natural history, navigating what was becoming at the time an increasingly tenuous balance between a nostalgic, romantic view of nature, and one mobilized by standardization, objectification, and the technologies of modernity. But as I suggested in the introduction, the text does more than provide us with insight into how nature and insects were "given meaning" at a particular historio-cultural moment—it reveals also how the meanings of nature were received by the author and her readership. If we do Budgen the posthumous courtesy of devoting to her text the kind of close, patient attention she encourages us to turn to its referent, what we find in our hands is not merely an attractive and quaint trilogy of books, but a portal to an elaborate, wondrous view of a living nature—a nature with a voice.

## NOTES

1. Gosse, *Romance of Natural History*, v.
2. Ibid.
3. See Crist, *Images of Animals*; and Crist, "'Walking on My Page.'"
4. Although the title pages of each consecutive volume are dated 1849, 1850, and 1851, respectively, the first volume appeared in December 1848, and the following two in each December thereafter, as the press reviews attest.
5. Crist, "Against the Social Construction of Nature and Wilderness," 8.
6. In the United States, the monochrome editions sold for two dollars each, the hand-colored for four dollars.
7. Cosslett, *Talking Animals in British Children's Fiction*, 1.
8. Ibid.
9. Budgen, *Episodes of Insect Life*, 1:xii. All subsequent instances of this source will be cited parenthetically in the text.
10. Bown, *Fairies in Nineteenth-Century Art and Literature*, 125–31.
11. Clark, *Bugs and the Victorians*.
12. Thomas, *Man and the Natural World*, 62.

## BIBLIOGRAPHY

Bown, Nicola. *Fairies in Nineteenth-Century Art and Literature*. Cambridge: Cambridge University Press, 2001.

Budgen, L. M. *Episodes of Insect Life*. 3 vols. London: Reeve, Benham, and Reeve, 1849–51.

Cary, Alice. *Clovernook; or, Recollections of Our Neighbourhood in the West*. New York: Refield, Clinton Hall, 1852.

Clark, J. F. M. *Bugs and the Victorians*. New Haven: Yale University Press, 2009.

Cosslett, Tess. *Talking Animals in British Children's Fiction, 1786–1914*. Aldershot, U.K.: Ashgate, 2006.

Crist, Eileen. "Against the Social Construction of Nature and Wilderness." *Environmental Ethics* 26, no. 1 (2004): 5–24.

———. *Images of Animals: Anthropomorphism and Animal Mind*. Philadelphia: Temple University Press, 1999.

———. "'Walking on My Page': Intimacy and Insight in Len Howard's Cottage of Birds." *Social Science Information* 45, no. 2 (2006): 179–208.

Gosse, Philip Henry. *The Romance of Natural History*. 6th ed. London: Nisbet, 1873.

Stewart, Susan. *On Longing: Narratives of the Miniature, the Gigantic, the Souvenir, the Collection*. Durham: Duke University Press, 1993.

Thomas, Keith. *Man and the Natural World: Changing Attitudes in England, 1500–1800*. London: Penguin, 1984.

**8**

Interacting with *The Watchful Grasshopper*; or, Why Live Animals
Matter in Twentieth-Century Science Museums

*Karen A. Rader*

Contemporary natural history museum scientists laud their institutions (without intentional irony) as "first and foremost, a celebration of what time has done to life"[1]—even though most nonhuman animals in their halls are dead, appearing only as stuffed or skeletal remains. More recently live animal displays have proliferated at these same museums, as their institutional boundaries have grown increasingly porous. Collecting and displaying live animals in museums began in European cabinets of curiosity,[2] and the practice became relatively common as early as the mid-nineteenth century throughout Europe. By the twentieth century, however, many American natural history museums and zoos featured similar (if not exactly the same) live animal demonstrations and displays.[3] For instance, the Bronx Zoo and the American Museum of Natural History both launched live butterfly exhibits in the 1990s, and the Smithsonian National Museum of Natural History still houses a popular "Insect Zoo," first opened in the 1970s.[4] More recently, when two of California's oldest cultural organizations—the San Diego Zoo and the San Diego Natural History Museum—contemplated a merger, zoo director Doug Myers portrayed the resulting mix of living animals and animal fossil remains as intrinsically complementary. "From our standpoint," he noted, "this is a natural. It completes the story that we tell."[5]

In 1973 the Exploratorium—a hands-on science center in San Francisco—developed an exhibit called *The Watchful Grasshopper* that complicates Myers's narrative of easily naturalizing live animals in museums. *The Watchful*

*Grasshopper* shared some display strategies and pedagogical goals with earlier live animal exhibits, so understanding its continuity with past practices reveals how live animals have been a part of museum visions throughout the twentieth century. But *The Watchful Grasshopper* display parted company with what came before it in one very important way: it asked visitors to engage with its animal subject by playing the role of a scientific experimenter. Visitor responses (preserved in the museum's archives) suggest that this approach backfired. Rather than teaching a simple scientific concept and drawing attention to the excitement of doing life science, as its designers had intended, embodied interaction with live animals through experimentation led museum visitors to reflect on the limits of laboratory methods of control and manipulation.

*The Watchful Grasshopper*, then, also illustrates the transgressive and contradictory possibilities of interactive live animal displays in museums. As museum studies scholars have long known, the creation of exhibits does not always map neatly onto how visitors consume them. In this case, one small living insect—a known agricultural pest, hooked up to electrodes, under a dome cover—mattered for Exploratorium visitors beyond the biological concept of perception through neural transmission it was meant to display. Instead of naturalizing relationships emblematic of the museum space—between dead and living animals, or between animal subjects and human scientists—*The Watchful Grasshopper* simultaneously revived visitors' interest in animal life and destabilized beliefs about how (and who) should best make natural knowledge from it.

## Prehistory: Live Animals in American Natural History and Science Museums, 1920s–1960s

Live animals in American museums go back at least to the 1860s.[6] But between 1920 and 1950, New York's American Museum of Natural History (AMNH) mounted displays that (in retrospect) epitomize two broad categories of "state of the art" exhibits with live animals by the middle of the twentieth century. First, in the 1920s, AMNH entomologist Frank Lutz envisioned introducing live animals into natural history museum displays as a corrective to boring dioramas. A masterful popularizer,[7] Lutz believed that static, "dead" displays (however beautiful) would not help make visitors want to learn more about entomology. "It is mighty difficult to make dead insects look happy on or under a sheet of celluloid water," he wrote. As early as the 1920s, Lutz expressed interest in building a habitat group of aquatic insects, but the underfunded entomology department could not compete for resources with other research. So Lutz took matters into his own hands: he put bowls of water with live aquatic insects and plants into exhibition cases. Encouraged by visitors' responses, Lutz next exhibited a wire cage of "trim, up-on-their-toes cockroaches," on which, he recalled, "even New Yorkers stopped to gaze." "The evolution of museum material which has been from the very dead to the almost life-like is not going to stop there but is going to take the next and apparently logical step," Lutz concluded.[8]

The displays of AMNH herpetologist Gladwyn Kingsley Noble, an early animal behavior scientist, continued this trend but toward slightly different ends. Noble displayed his research animals—including snakes, chameleons, and rats—which were collected from the wild and then bred in the museum for his Laboratory of Experimental Biology.[9] For Noble, a fellow curator later reflected, this "living material was added . . . to demonstrate particular points in natural history." In exhibits sponsored by Noble's department, live animals were displayed alongside devices designed to insure a dynamic presentation of key biological concepts.[10] For example, a 1940 AMNH *Annual Report* describes how the concept of protective coloration was illustrated in a display of live copperhead snakes, which blended into the background of the exhibit painted by Works Progress Administration artists. The function of such blending was highlighted by "a mechanical device in which a model snake is made to appear and disappear alternately."[11] Another mid-century AMNH exhibit on the idea of positive reinforcement included rats finding their way through a maze to food. This rodent display was mounted alongside an adjacent, larger-scale maze, designed for the human visitors to navigate and then compare their performance (with no food incentive) to that of the animals.

Starting in the early 1950s, science museums introduced yet another approach: live animal exhibits that simultaneously took both biological concepts and human-animal interactions as their subjects (even when those subjects presented contradictions). The chick hatchery display at the Chicago Museum of Science and Industry represents a case in point. First mounted as a temporary display in April 1953, to attract visitors during the Easter season, Chicago curators installed a large, glass-topped incubator in the International Harvester tractor exhibit's farm. This allowed visitors to observe baby chicks developing and hatching.[12] The chick hatchery attracted so many visitors that Swift Nutrition, an international food corporation, approached the museum later that summer about "an extensive exhibit on the field of nutrition." The resulting Food for Life Hall displayed (according to in-house descriptions) the "complete story of man's food: where it comes from, how it is adapted to fit his needs, how the various elements entering into the production of food affect man." By the end of the decade, the success of the chick hatchery spawned a full-scale animal nursery at the museum. Visitors could feed a multitude of agricultural food animals, including piglets, lambs, calves, rabbits, and duck-lings,[13] and follow their favorite animals in the museum's member newsletter, where the education department reported on their developmental progress.[14]

The increased presence of all kinds of live animal exhibits in both natural history and science museums probably played some role in shifting visitor expectations of displays—and of museums in general. Visitors to the Smith-sonian National Museum of Natural History and Boston Museum of Science, for example, complained about boring halls and lifeless exhibits. In Boston, some of these displays remained from its earlier incarnation as the Boston Society of Natural History Museum. In late 1956, Boston's director, Bradford Washburn, received a letter from "Mrs. Cannon," a self-described friend of a trustee, who lamented that existing hands-on microscope displays were inadequate. The public had an interest, Cannon argued, in seeing the living microscopic world, not dead specimens under slides.[15] "We are intent on giv-ing a new dimension to the word Museum," Washburn promised his trustees that same year. "Our Museum of Science is primarily a teaching institution," he argued, noting that its education staff was double that of exhibits, and that science teachers, not research specialists, predominated in most newer muse-ums' staff. New education department members, he concluded, "were selected for the ability to create and sustain an enthusiasm for learning. Our exhibits, too, are primarily designed for teaching and the majority of them demand the *active participation of the visitor* in the learning process."[16]

Perhaps not surprisingly, then, displays featuring live animals soon moved further beyond dioramas and enclosed cases, to allow visitors to engage more directly with a single (or several) star animals. The Boston Museum of Sci-ence, for instance, featured what exhibit designers dubbed "animal demon-strations," which entailed carefully monitored, hands-on contact between visitors and living animals. Museum education department staffers mediated these interactions: they invited children to touch animals like Black Beauty, a seven-foot-long indigo snake with unique camouflage markings, or the slightly more approachable Herkemiah and Cuddles, two pet porcupines whose quill-release mechanisms had been disabled.[17]

What became the Boston Museum of Science's most famous live animal demonstration arose by chance. After finding a fledgling owl and reviving the fluffy ball with a medicine dropper, two suburban Boston residents had called Washburn to see if the museum wanted the tiny bird. Christened "Spooky," the young owl was featured in exhibit halls after only a few weeks at the museum, and to museum staff, the owl took on near-human status as a charismatic educator. Chauffeured by assistant director of education Gilbert Merrill and gawked at by nearby drivers amazed to see a great horned owl perched on the front seat of a sedan, Spooky frequently visited New England schools and community groups, chalking up nearly 5,200 miles of road travel in a single year. By the mid-1950s, Spooky had become a celebrity outside the museum's walls, appearing in *Look* magazine and on NBC's *Today Show*.[18]

Washburn claimed that Spooky and other live animal demonstrations represented new educational successes because they were hands-on. As he later recalled, "The idea was, if you mix this [interactivity] with natural history and just simply say 'we learn in this place, everything is fun to learn,' young people would be interested—because . . . they were discovering things, whatever they might be."[19] But, in fact, such displays encouraged younger visitors' interactivity and engagement with science only abstractly. The *Museum of Science Newsletter* reports suggest that Spooky's initial stillness "engaged" schoolchildren primarily by raising in them the question: Is this a live animal or a stuffed specimen? Most visitors seem (from extant archival photos) to have stared at the animal and puzzled over Spooky's tameness.[20]

Later in the decade, the Boston Museum developed an arrangement that allowed more direct animal-visitor interaction: they hired teenaged "junior assistants" who were charged with caring for the museum's growing menagerie. High school students could get behind the scenes and learn about the animals—not only their natural history and behavior but also what it took to keep them healthy and happy for museum use. Washburn argued that these students were participating in "science in action"—and he featured the junior assistants in a museum newsletter story about educational outreach.[21] But by then it was too late.[22]

## Animal Behavior at the Exploratorium: Developing *The Watchful Grasshopper*

When the Exploratorium was founded in 1969, its exhibit designers took "science in action" to a new level. Director Frank Oppenheimer envisioned a series of exhibits on perception that would demonstrate commonalities in the way plants, animals, and human beings perceived stimuli; these would complement the successful hands-on physics exhibits that formed the core of the Exploratorium's existing displays. Oppenheimer hesitated before building life science displays, because he was worried that live plants and animals would transform the Exploratorium into a zoo or a botanical garden or, worst of all, into a traditional museum that housed specimen collections. Living specimens demanded ongoing care; they would require the Exploratorium to hire trained biologists, whose interests and skills would likely be entirely different from those of the freewheeling exhibit builders he had wanted to court.[23] Ultimately, the National Science Foundation put significant pressure on the Exploratorium to include biology in its exhibits, and also offered seed money, so the project was launched in 1972.[24]

Oppenheimer hired Evelyn Shaw as the Exploratorium's first curator of life science. Shaw had extensive experience with live animals in museums: she worked in Gladwyn Kingsley Noble's old department (now renamed Animal Behavior) at the American Museum of Natural History. Oppenheimer offered her a chance to continue her scientific work (on marine animal biology) and to publicize her methods and results through exhibits in the Exploratorium.[25] In 1972, Shaw hired a young Berkeley alumnus, Charlie Carlson, to assist her. Carlson was, temperamentally, a perfect fit: he was comfortable with interdisciplinarity, experimentation, and idealism—he had double-majored in zoology and communications and embraced the creative culture of 1960s Berkeley—but he was also pragmatic when it came to making scientific concepts meaningful.[26]

Throughout the next several years, Shaw and Carlson together created a series of interactive displays themed around animal behavior. Their work drew inspiration from earlier exhibit strategies, but also developed a new mode of visitor-animal interaction. Unlike those earlier live animal displays in natural history and science museums, Shaw and Carlson's exhibits demanded more of visitors than just observing or animal-petting: they put the visitor in the role of an experimental scientist. Their displays (like Noble's and Lutz's) placed live animals "under glass" (so that visitors were not actually touching them), but unlike with live animal demonstrations, they relocated control of the animal-visitor interaction—away from the exhibit designer or the educational demonstrator, and toward the visitor.[27]

The exhibit best illustrating Shaw and Carlson's approach was *The Watchful Grasshopper*, which featured a live grasshopper, under a dome, with wire electrodes inserted into its ventral nerve cord (fig. 8.1). This procedure, explained the signage, did not make the grasshopper uncomfortable and allowed visitors

FIG. 8.1 Sketch of *The Watchful Grasshopper* exhibit as it appeared in the *Exploratorium Cookbook II*. Photo © Exploratorium (http://www.exploratorium.edu/).

to explore the grasshopper's visual field, in order to determine what triggered impulses in the insect. The electrodes were hooked up to measuring devices that were also a part of the display: an oscilloscope, which recorded extracellular signals in the grasshopper's brain, and amplifying speakers, which allowed visitors to hear, not merely see, the oscilloscope's activity as it "clicks." Shaw and Carlson imagined that when visitors moved in front of the animal's visual field, they could watch and hear the neural effects of the grasshopper watching them.

The relative sophistication of the grasshopper exhibit setup, combined with the fact that it featured a live animal, presented significant challenges for its creation. First, although Carlson and Shaw settled quickly on *Schistocerca nitens* as the insect that would most clearly illustrate the relationship between neurological stimuli and animal behavior for visitors, its status as an agricultural pest in California meant that the insects were nearly impossible to acquire through commercial venues. This forced staff members into temporary careers in grasshopper husbandry, breeding whatever grasshoppers would be used in the display. Exhibit builders promptly incorporated their newly acquired knowledge of grasshopper husbandry and anatomy into the exhibit. Carlson created a supplementary display explaining the grasshoppers' life cycle, featuring the museum's grasshopper colony and the exhibit designers' scientific knowledge of behavior and husbandry.[28]

The exhibit also required museum staff to become experts in the frustrating field of grasshopper surgery. Carlson read what he could on *S. nitens*, conducting his own intensive observations and experiments to learn how to insert electrodes into their ventral nerve cords; then he taught his exhibit support staff to do the same. This required considerable practice and skill. Grasshoppers were immobilized with dental wax—if they moved too much, they were anesthetized through ten minutes of refrigeration—then placed under a dissecting microscope, and the electrode implanted and tested. Finally, the preparator used a mixture of beeswax and rosin to cement the electrodes inside the animals. "If this preparation is carefully made without too much trauma to the animal," Carlson wrote, "it will last one week or more, up to a month, without noticeably affecting the behavior or health of the animal," a period of time that already overextended staff members must have found short.[29] Graphics that accompanied *The Watchful Grasshopper* went to great lengths to illustrate the electrode implantation procedure, also emphasizing that it did not hurt or permanently injure the insects.[30]

### Interacting with *The Watchful Grasshopper*: Visitor Perspectives

When it was first mounted, biologists and fellow museum workers hailed *The Watchful Grasshopper* and other similar life science displays as a "tour de force of exhibit construction."[31] After Carlson and his staff traveled to the 1977 meeting of the American Neurobiological Association to demonstrate the displays to academic scientists, for example, he received numerous requests to create instructions so university teachers could replicate *The Watchful Grasshopper* and a host of other displays that the Exploratorium had regrouped into a broader exhibition, *The Language of the Nerve Cells*. Local medical school physiology professors brought their classes to use the Exploratorium biology displays as laboratory experiments. Biology graduate students suggested additions that would make the exhibits more useful for their own research.[32]

But although popular with scientists, these displays did not draw the same response from the broader public: according to an in-house study only about

5 percent of the Exploratorium's 560,000 annual visitors interacted with the displays.[33] Further, visitors drew conclusions about science and animal behavior from the museum's exhibits, but these conclusions did not always echo those drawn by the Exploratorium's own staff and the broader scientific community. In 1981, for example, a local teacher ("M. Clausen") lauded the Exploratorium as a "wonderful place for making people become aware of and excited by science," but noted that she found *The Watchful Grasshopper* downright disturbing. "Whatever the instructional value of such an exhibit it represents cruelty to animals and only encourages people to treat animals as playthings without feeling," she wrote to Oppenheimer. "I sincerely hope you will remove that torture chamber. I did not go through the animal behavior section after seeing the grasshopper in fear of seeing more of such disturbing sights."[34] Perhaps the museum could show a short animated film with sound to illustrate this phenomenon instead, she suggested.

Another visitor, Leelane Hines, also registered his discomfort with the exhibit, describing it as "counterproductive" and noting that "what we learn is not always what people think they are teaching." Rather than conveying information about animal behavior and neurological pathways, Hines argued, the exhibit propagated a reprehensible lesson: "When creatures are less than human, we, as superior more knowledgeable beings, need not treat them with respect or kindness. Lesser beings may freely be used for our own (scientific) (genetics) (self-defense) purposes."[35]

Oppenheimer himself sometimes responded directly to these concerns, by explaining both the scientific methods and motivations behind Shaw and Carlson's life science displays, especially to a number of visitors who championed animal rights and found the museum's experimentation on live specimens suspect as a result. In 1981, for example, he attempted to reassure an angry Henrietta Gennrich that the grasshoppers in the exhibit were not receiving visitor-administered electrical shocks. "The live grasshopper that you saw was not hooked up to a shock stimulator!" he protested. "Fine, flexible wires were connected to the back of the grasshopper who lives very happily and can move around," but measurement was very different from torture, he explained. Still, he agreed, the exhibit could probably use some clarification. Going forward, he promised, "we will do everything we can to make it clear that we are showing that animals, as well as people, transmit information by generating their own electricity and that this process is going on all the time in all of us."[36]

As a matter of pedagogical philosophy, Exploratorium exhibits intentionally excluded broader political and cultural discussions of science and its methods. Oppenheimer actively disavowed exhibits whose lessons could be easily applied to social issues. Because he believed that the educational mission of his reformed museum "was not to give people the right answers . . . but to help them gain the confidence to make discoveries for themselves," he resisted staff and visitor calls to develop exhibits promoting environmental or popular cultural interests.[37] "We have to develop new tools to persuade people to act sensibly," Oppenheimer told a reporter in 1979, and "we don't have to rely on coercion. That is the meaning of a free society."[38] Scientists who believed their work to be largely apolitical admired this stance (and understood it to have come from Oppenheimer's own life experience),[39] but museum publics—visitors and science educators—saw it as either naive or misguided. Science, and especially life science, was necessarily political, they argued, and as such, science education should place scientific developments and discoveries in social and cultural context.[40] Others outside the museum community disagreed. The Exploratorium was right to maintain a careful innocence, Lexington, Kentucky, reporter Walter Sullivan editorialized, for this was the only way it could demonstrate a "clean" message: "In science, there is also beauty and joy."[41]

Those visitors committed to the expansion of animal rights were unconvinced by Oppenheimer's reasoning, and not easily converted to a purely rationalist perspective. In the case of *The Watchful Grasshopper* (unlike Noble's rat and human maze, exhibited in the 1930s), the comparison between animals and humans generated controversy and actively disrupted the museum's pedagogical mission. Visitor Muhaima Startt made this point explicit: she insisted that the Exploratorium would never conduct a similar experiment on a human being, and, as a result, that the museum's life science exhibits were fundamentally incompatible with a broader respect for life in all its forms. "My simple point of view is that animals have their own things to do in their natural environment, and that should be respected," Startt wrote. If visitors wanted to experiment on something, she concluded, they would do better to experiment on themselves: "It may be of more learning to folks to have more tools for self-exploration available—the bicycle reading one's heartbeat and the simple EMG register are well thought through in this respect."[42] Ultimately, many Exploratorium visitors concluded that the museum prioritized scientific discovery over the preservation of animal life—even when the animal life in question was not a charismatic mega-fauna but rather belonged to an aggressive agricultural pest insect.

## Conclusion

For the historian, then, live animal displays represent a kind of sampling device for shifting attitudes toward animals, science, and museums in mid-twentieth-century America. The insect subject is often figured as somehow "not an animal" or "not as much an animal" as vertebrates. Indeed, mid-twentieth-century visitors to the Exploratorium would have already been inundated with film and advertising portrayals of insects as machinelike, even militaristic.[43] But as *The Watchful Grasshopper* and this volume's other insect examples (Ogilvie's painted butterflies and Dodd's anthropomorphized crickets) reveal, ethical and empathetic attitudes toward insects—even those whose bodies are "coupled" with electronic hardware—can be recovered only if we look beyond these displays, to how humans actively engage with them.

Over the course of the twentieth century, museums themselves have changed as much as the use of live animals in their exhibits has. As museum missions moved more toward public education, live animals as mere presences—as dynamic reflections of textbook concepts or museum research programs—gave way to modes of display that laid claim to different kinds of educational expertise: first, through controlled live demonstrations, and later, through exhibits in the Exploratorium where the museum visitor was called on to directly manipulate animals. All these new displays were difficult to create and maintain—just as controlling living nature in laboratories or other domesticated spaces is[44]—but that, in fact, was the point. The subject of these newer displays was never intended to be the animals (as it was in some earlier exhibits created by zoos and aquaria);[45] rather, the subject was "science in action." Authenticity, to the extent that it was achieved by *The Watchful Grasshopper*, was measured not according to the animal subjects' life on display (as with Thorsen's Saint Bernard Barry) but, rather, according to whether the display's instrumentation (wires, oscilloscope, sound clickers) accurately educated the human visitor about the neurology of vision.

But just as in the early modern period, with Ruud's monstrous pigs, museums' complex web of stakeholders ensured that the scientific meaning of this exhibit could not be so easily contained by its intentional design. In *The Watchful Grasshopper*, nonscientist visitors saw the contradictions, rather than the naturalness, of live animals in the museum space. By presenting a live insect and engaging the visitor in an experiment with it, *The Watchful Grasshopper* enabled the animal to be viewed both as a holistic form of life and as an analytic tool of scientific knowledge-making. Exhibit designers, without intending to introduce a discussion of "the co-constitutive, entangled, responsible, and responsive relationships that we might form outside of and against" laboratory science's relations with animals,[46] nevertheless evoked this response in nonscientist visitors by compelling them to become scientists. Perhaps it is not surprising, then, that this extraordinary display had a relatively short life on the science museum floor. Although it was included in the Exploratorium's second "cookbook" of exhibits, there is little evidence that other museums reproduced or otherwise emulated the display—and it was removed from the Exploratorium sometime in the mid-1980s. Its legacy, however, remains—in persistent debates among museum, zoo, and aquaria visitors and scientists about the role of natural knowledge and scientific techniques involved in museum and zoo displays.[47] That such debates remain alive and well suggests that animal exhibits (even if not dead and stuffed) still have something important to tell us about the public understanding of science—and life—in museums.

## NOTES

1. See, for example, Fortey, *Dry Storeroom No. 1*, 24.
2. George, "Alive or Dead."
3. Hanson, *Animal Attractions*; Rothfels, *Savages and Beasts*. See also Kohlstedt, "Entrepreneurs and Intellectuals."
4. Wade, "Winged Messengers from Fragile Forests"; Martin, "Exhibits Aflutter at the Bronx Zoo"; Casey, "Randy Roaches, Dancing Bees."
5. Myers is quoted in Steele, "Similar Missions Driving Zoo, Museum Unity Talks." This particular merger never came to fruition.
6. The Smithsonian Institution, for example, housed a live beehive exhibit (in a glass case, as part of an exhibit on agriculture and food) starting in the 1890s. See Smithsonian Institution Archives Record Unit 285, Box 16, Folder 4—with thanks to Pamela Henson for drawing my attention to this display. Also, P. T. Barnum's rebuilt American Museum displayed live boa constrictors eating live rabbits, as part of what Barnum told animal anticruelty advocate Henry Bergh was a commitment to showing "animals here as nearly in their natural state as they can be exhibited." See correspondence between Barnum and Bergh (along with Barnum's letter of support from Louis Agassiz) at "The Lost Museum," Digital History Archive, http://chnm.gmu.edu/lostmuseum/lm/192/ (accessed February 28, 2011).
7. Lutz quoted in Barton, "Attorney for the Insects," 182.
8. Lutz, "Use of Live Material in Museum Work," 8.
9. Two excellent secondary accounts of Noble's work are Mitman, "When Nature 'Is' the Zoo"; and Milam, *Looking for a Few Good Males*, 62–67.

10. Myers, "History of Herpetology at the American Museum of Natural History." See especially the section titled "The Museum as Zoo" (pp. 87–88), which reflects Myers's own prejudice more than the actual preponderance of live animal displays at the American Museum of Natural History.

11. *American Museum of Natural History Annual Report, 1940*, 17–18; as of this publication, all annual reports for AMNH are electronically accessible at http://digitallibrary .amnh.org/museum/annual_reports/about. Earlier AMNH displays featured "exotic" specimens—such as in 1936, when the museum displayed the "only living specimens of the Matecumbe Chicken Snake, a form recently destroyed by tornadoes in Florida." See *American Museum of Natural History Annual Report, 1936*, 12.

12. "Visitors from Around the Globe: Easter Crowds Watch Chickens Hatch on Harvester Farm," *Progress* 4, no. 3 (May/June 1953): 8, Chicago Museum of Science Archives (hereafter CMOSIA).

13. "Swift Moves In," *Progress* 4, no. 5 (September/October 1953): 2–3, CMOSIA.

14. See, for instance, the picture of the chicks in *Progress* 9, no. 2 (March/April 1958): 7; and "Baby Chicks Make Easter Perfect," *Progress* 9, no. 3 (May/June 1958): 5 (copy filed in "Swift Nutrition exhibit" folder)—all CMOSIA.

15. "Memo: Mrs. Cannon's Letter," Bradford Washburn to Norm Harris, December 31, 1956, Folder "C," Bradford Washburn Papers, Boston Museum of Science Archive (hereafter BMOSA).

16. *Boston Museum of Science Annual Report, 1955–56*, 3, BMOSA.

17. Bradford Washburn Oral History conducted by the author and Sylvie Gassaway, 2005 (transcript in author's personal possession); see also *Boston Museum of Science Annual Report, 1950*, BMOSA; and Rock, *Museum of Science, Boston*, 101.

18. Rock, *Museum of Science, Boston*, 104.

19. *Progress Report: Museum of Science* 6 (May 1950): 1, BMOSA; see also Washburn's report on his trip to European museums during the summer of 1957, as quoted in "Minutes of Museum of Science Trustee Annual Meeting, 9 October 1957," 5, in Box "1957–58" of the Washburn Papers, BMOSA.

20. See, for example, the feature articles and pictures of Spooky in the *Museum of Science Newsletters* of 1954 and 1959, BMOSA.

21. "Junior Assistants in the Animal Room," *Museum of Science Newsletter*, March 1956; the "science in action" quote is from the March 1957 issue of the *Newsletter*—both in BMOSA.

22. See, for example, correspondence pertaining to the failure of the Boston Museum of Science's approach in the eyes of the U.S. National Science Foundation (NSF): Bradford Washburn (BW) to George Rothwell (NSF program officer), December 7, 1961, Box "1961," Folder "Correspondence"; "Memorandum—Visit by BW to NSF, 14 November 1962," Box "1962," Folder "NSF"; see also Norm Harris to BW, January 6, 1959, memo regarding "Summer Teachers Courses," Box "4-1959," Folder "Education Dept."—all from Washburn Papers, BMOSA.

23. Hein, *The Exploratorium*, 92.

24. Oral History interview with Charles Carlson by author and Sylvie Gassaway (2005); transcript of this interview is in the author's possession. See also Hein, *The Exploratorium*, 91, 93; and the Exploratorium's NSF application, Exploratorium Records, BANC 87/148c, 35:17, Bancroft Library, University of California–Berkeley (hereafter BL-UCB). Earlier NSF funding for science education at the Exploratorium had been provided through the section on Scientific Curriculum Improvement; see James Strickland to Frank Oppenheimer, January 20, 1971, BANC MSS. 87/148c, 7:44, BL-UCB.

25. Cahn and Shaw, "Method for Studying Lateral Line Cupular Bending in Juvenile Fishes"; see also Oral History Interview with Charles Carlson (2005).

26. As cited in Cole, *Something Incredibly Wonderful Happens*, 167.

27. *The Brine Shrimp Ballet*, another such exhibit on which Shaw and Carlson worked, is outlined in Hipschman and the Exploratorium staff, *Exploratorium Cookbook II*, recipe #99.

28. These additional *Watchful Grasshopper* graphics are described in ibid., recipe #124.

29. Carlson describes the display setup in Carlson et al., "Two Simple Electrophysiological Preparations Involving the Grasshopper," 182.

30. See Hipschman and the Exploratorium staff, *Exploratorium Cookbook II*, recipe #124.

31. Hein, *The Exploratorium*, 107.

32. For example, see Frances Clark to Frank Oppenheimer, June 29, 1979, BANC MSS. 87/148c, 4:4, BL-UCB; and the letter from two teaching assistants in the Department of Applied Behavioral Sciences at UC Davis to Frank Oppenheimer, April 7, 1977, detailing their students' successful "experiential learning" at the Exploratorium, available at BANC MSS. 87/148c, 4:4, BL-UCB.

33. Carlson et al., "Report to Grass Foundation" (1977), BANC MSS. 87/148c, 35:38, BL-UCB; see also an earlier in-house study in the archives, "Documentation of Exhibits, the Exploratorium, December 1974 to June 1975," BANC 87/148c, 39:8, BL-UCB.

34. M. [illegible name] Clausen (who identified herself as a teacher) to "Director of the Exploratorium," October 1981, BANC MSS. 87/148c, 4:4, BL-UCB.

35. Leelane E. Hines to Frank Oppenheimer, BANC MSS. 87/148c, 4:13, BL-UCB.

36. Frank Oppenheimer to Henrietta Gennrich, dated March 1981, BANC MSS. 87/148c, 3:18, BL-UCB.

37. See Theodore Sudia to Frank Oppenheimer, November 19, 1975, which suggests that Oppenheimer consider some exhibits under development with the "Man and the Biosphere" program of UNESCO. BANC MSS. 87/148c, 4:23, BL-UCB.

38. See unpublished essay "Science and the Ethics or Coercion," and letter from Oppenheimer to K. C. Cole, BANC MSS. 87/148c, 21:32, BL-UCB.

39. Oppenheimer was accused of communist associations by HUAC, and in 1957, a few months shy of tenure, he left the University of Minnesota—and academic physics—to become a cattle rancher in Pagosa Springs, Colorado. While there, a local high school learned of his presence and asked him to teach science classes—and in this way, Oppenheimer returned to science via science education. On Oppenheimer's biography, see Cole, *Something Incredibly Wonderful Happens*, part 1.

40. Hein, *The Exploratorium*, 109. Hein also notes that Oppenheimer and Carlson were both "summoned to debates" on local radio and television programs to defend "The Watchful Grasshopper" from animal rights critics (p. 103).

41. Sullivan, "Science Centers Seek Public Participation."

42. Muhaima Startt to "Executive Director, Exploratorium," July 28, 1984, BANC MSS. 87/148c, 4:23, BL-UCB.

43. See Palladino, "Ecological Theory and Pest Control Practice"; Zerner, "Stealth Nature"; Buhs, *Fire Ant Wars*. For an analysis of this phenomenon that looks toward the twenty-first century, see Dodd, "Trouble with Insect Cyborgs."

44. See Rader, *Making Mice*.

45. See Hanson, *Animal Attractions*; Rothfels, *Savages and Beasts*. Debates over the educational value of such animal displays have recently reignited; see Marino et al., "Do Zoos

and Aquariums Promote Attitude Changes in Visitors?"; and the response, Falk et al., "Critique of a Critique."

46. Weisberg, "Broken Promises of Monsters," 58.

47. Rader and Cain, *Life on Display*, chapter 7.

## BIBLIOGRAPHY

Barton, D. R. "Attorney for the Insects." *Natural History* 48 (1941): 181–85.

Buhs, Joshua. *The Fire Ant Wars: Nature, Science, and Public Policy in Twentieth-Century America*. Chicago: University of Chicago Press, 2005.

Cahn, Phyllis, and Evelyn Shaw. "A Method for Studying Lateral Line Cupular Bending in Juvenile Fishes." *Bulletin of Marine Science* 15, no. 4 (1965): 1060–71.

Carlson, Charles, Bonnie Jones, and Wayne La Rochelle, with illustrations and photographs by Susan Schwartzenberg. "Two Simple Electrophysiological Preparations Involving the Grasshopper." *Association for Biology Laboratory Education Manual* 2 (1980): 181–95.

Casey, Phil. "Randy Roaches, Dancing Bees: The Insect Zoo Story." *Washington Post*, August 1, 1971.

Cole, K. C. *Something Incredibly Wonderful Happens: Frank Oppenheimer and the World He Made Up*. Boston: Houghton Mifflin Harcourt, 2009.

Dodd, Adam. "The Trouble with Insect Cyborgs." *Society and Animals*, forthcoming.

Falk, John H., Joe E. Heimlich, Cynthia L. Vernon, and Kerry Bronnenkant. "Critique of a Critique." *Society and Animals* 18 (2010): 415–19.

Fortey, Richard. *Dry Storeroom No. 1: The Secret Life of the Natural History Museum*. New York: Knopf, 2008.

George, Wilma. "Alive or Dead: Zoological Collections in the 17th Century." In *The Origins of Museums*, edited by Oliver Impey and Arthur MacGregor, 179–87. Oxford: Clarendon Press, 1985.

Hanson, Elizabeth. *Animal Attractions: Nature on Display in American Zoos*. Princeton: Princeton University Press, 2002.

Hein, Hilde. *The Exploratorium: Museum as Laboratory*. Washington, D.C.: Smithsonian Institution Press, 1990.

Hipschman, Ron, and the Exploratorium staff. *Exploratorium Cookbook II: A Construction Manual for Exploratorium Exhibits*. San Francisco: The Exploratorium, 1983.

Kohlstedt, Sally G. "Entrepreneurs and Intellectuals: Natural History in Early American Museums." In *Mermaids, Mummies, and Mastodons: The Emergence of the American Museum*, edited by William Alderson, 23–39. Washington, D.C.: American Association of Museums.

Lutz, Frank E. "The Use of Live Material in Museum Work." *Museum News* 6 (1930): 7–9.

Marino, Lori, Scott Lilienfeld, Randy Malamud, Nathan Nobis, and Ron Broglio. "Do Zoos and Aquariums Promote Attitude Change? A Critical Evaluation of the American Zoo and Aquarium Study." *Society and Animals* 18 (2010): 126–38.

Martin, Douglas. "Exhibits Aflutter at the Bronx Zoo." *New York Times*, May 23, 1996.

Milam, Erika. *Looking for a Few Good Males: Female Choice in Evolutionary Biology*. Baltimore: Johns Hopkins University Press, 2010.

Mitman, Gregg. "When Nature 'Is' the Zoo: Vision and Power in the Science of Natural History." *Osiris* 11 (1996): 117–43.

Myers, Charles W. "A History of Herpetology at the American Museum of Natural History." *Bulletin of the American Museum of Natural History* 252 (2000): 1–231.

Palladino, Paolo. "Ecological Theory and Pest Control Practice: A Study of the Institutional and Conceptual Dimensions of a Scientific Debate." *Social Studies of Science* 20, no. 2 (1990): 255–81.

Rader, Karen A. *Making Mice: Standardizing Animals for American Biomedical Research, 1900–1955.* Princeton: Princeton University Press, 2004.

Rader, Karen A., and Victoria E. M. Cain. *Life on Display: Revolutionizing Museums of Science and Nature, 1900–1990.* Chicago: University of Chicago Press, forthcoming.

Rock, Mary Desmond. *The Museum of Science, Boston: The Founding and Formative Years: The Washburn Era, 1939–1980.* Boston: Boston Museum of Science, 1989.

Rothfels, Nigel. *Savages and Beasts: The Birth of the Modern Zoo.* Baltimore: Johns Hopkins University Press, 2002.

Steele, Jeanette. "Similar Missions Driving Zoo, Museum Unity Talks." *San Diego Union-Tribune,* May 18, 2010.

Sullivan, Walter. "Science Centers Seek Public Participation." *Lexington (Ky.) Dispatch,* September 17, 1975.

Wade, Nicholas. "Winged Messengers from Fragile Forests." *New York Times,* November 27, 1998.

Weisberg, Zipporah. "The Broken Promises of Monsters: Haraway, Animals, and the Humanist Legacy." *Journal of Critical Animal Studies* 7, no. 2 (2009): 1–61.

Zerner, Charles. "Stealth Nature: Biomemesis and the Weaponization of Life." In *In the Name of Humanity: The Government of Threat and Care,* edited by Ilana Feldman and Miriam Ticktin, 290–324. Durham: Duke University Press, 2010.

## Polar Bear Knut and His Blog

*Guro Flinterud*

Early in December 2006, at the Berlin Zoo, staff were preparing to receive a litter of polar bear cubs. A retired circus bear, Tosca, was now pregnant for the second time. Tosca had been unable to take care of her first cub, who consequently died. The keepers were anxious; would she be able to care for her cubs this time, or would she, like so many captive polar bears, abandon them after birth? On December 5, Tosca gave birth to two cubs. For a few hours she attempted to feed them, but eventually gave up, walked away, and went to sleep, leaving the cubs outside her maternity den. Zookeeper Thomas Dörflein and veterinarian André Schüle waited, hoping Tosca might return. But eventually they realized the zoo staff would have to take on the responsibility themselves, and hoisted the two cubs out of the enclosure with a small landing net.

Of the two cubs, one died after only a few days, but the surviving cub would become the most famous polar bear in Germany, if not the whole world: Polar Bear Knut (fig. 9.1). Knut's level of fame was unanticipated even by the zoo management, and should be seen as the result of a set of fortunate coincidences. He was born at the right time in the right place; with technologies of mass communication reaching new highs along with the ice in the Arctic reaching new lows, the Western world was more than receptive of a fluffy little polar bear successfully saved by human beings. In hindsight, Knut's highly publicized sudden death at age four in 2011 cast the story in a different light, foregrounding the problems concerning human interference in nature that had been all but suppressed in the early narrative.

FIG. 9.1 Polar bear Knut. Courtesy of Christina M. / cute-crazy-Knut blog.

Knut's celebrity is evidenced in many different places, but one of the most significant arenas in which he became an international star was the Internet. In March 2007, a little over three months after Knut's birth, the local broad-casting company Rundfunk Berlin-Brandenburg (RBB) started a blog that was supposedly written by Knut himself. In this essay, I look at how Knut's fans communicated within the blog. Originally planned as a place free of opin-ion, the many heated discussions that ensued reveal that the "story of Knut" carried with it connotations far exceeding the one-dimensional representation of a cute cub around which the blog was built. What can the breakdowns and negotiations within this online fan community teach us about the values and opinions connected to the polar bear in the early twenty-first century?

## Knut and the Blog

Knut's career as a blogger started on March 6, 2007, when the director of the RBB online department, Torsten Rupprich, envisioned creating a blog in which Knut would "write about" his own life. He described it as a *Schnapsidee*, as something of a whim—though a whim that would spawn a blog that at its most popular was one of the most visited sites in Germany.[1] The blog ran for almost two years, and was suspended in January 2009.[2]

RBB's sponsorship agreement with the Berlin Zoo gave them exclusive media access behind the scenes at the zoo in the months before Knut's first public appearance on March 23, 2007. Being a local company, RBB's TV programs were available only to viewers in the Berlin-Brandenburg area. In order to reach a broader audience, they made a site for Knut on their web page where videos of the cub were available as podcasts. The blog grew from this initial site, and was initially written by Torsten Rupprich. Swamped by the amount of work involved, Rupprich soon handed over the responsibility to another journalist at the online department, before eventually realizing that it would be best to hire someone from outside to do the job. In September 2007, writer Dorotea Horedt was hired, and she continued until the blog was stopped in January 2009. For a short period in late summer 2007, two of the regular visitors to the blog also had writing privileges, though they wrote mainly as themselves and not as Knut.

The blog quickly became immensely popular in Germany. During the first week that Knut appeared in front of the public at the zoo (the last week of March 2007), the blog was attracting almost fifty thousand hits daily. The comment count was also growing steadily, with around one hundred comments to every entry. By early May, the average was between two hundred and three hundred comments to each entry, with occasional entries receiving as many as seven hundred or eight hundred. What these figures reflect is not so much an increase in the number of visitors, but a solidifying of the community of returning visitors, resulting in lively discussions in the commentary sections of each entry. From the end of March onward, there was also an increase in the number of comments from people outside of Germany. Soon there were as many comments in English as in German, and a demand for translations arose. It started with a German fan who began translating the main entries into Spanish to help her improve her language skills. By May 2007, translations appeared in Spanish, English, Italian, French, Portuguese, Dutch, Polish, Hungarian, Korean, Chinese, Greek, and Finnish, all made by the fans.

The fan community in Knut's blog comprised people from a broad range of countries all over the world.[3] It is difficult to formulate a description of them as a group, mainly because the format does not encourage them to reveal personal details about themselves. Writer Dorothea Horedt also noted the plurality of the group. She met with several of the fans during the existence of the blog, and she stressed that she found it difficult to give a general description of "the typical Knut fan."

"Knut's" blog posts were a combination of text and photos or videos. The texts often commented on the scenes shown in the videos and photos, providing a translation from animal behavior to human experience. Thus the way the blog posts were composed, displaying Knut as both a living polar bear cub and a blogger, created a sense of increasing proximity of the human to the animal. "Knut" wrote about how he felt and what he thought, explaining his actions in a way typical of the child narrator of a children's story. In the very first blog entry he wrote about growing up as a polar bear in a zoo: "'Papa Thomas' and I are always practicing new stuff, so that I will become a real, grown polar bear one day. Here you can see me taking my first bath. Weee, I'm splashing around and 'Papa Thomas' says I'm already doing very well."[4] Consequently, when the fans communicated with Knut in the blog, they related to something that was not quite animal, yet not quite human. This blurring of boundaries between what is human and what is animal enabled the fans to address Knut directly as a human being, yet to still keep his animal identity active and acknowledged in their communications, where it was always clear that they were ultimately referring to a nonhuman animal. The balance of text and visuals seemed to display Knut as "human enough" for his fans to communicate with, yet "animal enough" to ensure they were able to maintain that their discussions were always about an animal.

The way Knut was represented in the blog could clearly be described as anthropomorphic. The stories in the blog were not merely *about* Knut; they were first-person accounts "written by" Knut, and the fans elected to address him directly. As such, Knut in the blog was both a representative of the species *Ursus maritimus* and an individual with his own personality and life history. When commenting on the content of the blog entries, the fans addressed their comments or questions to Knut, while addressing the actual writers as "administrators." The communication in the blog, however, went beyond just writing to Knut. The conversations among the fans in the commentary sections were just as important, and to understand the nature of the communication in the blog, one has to look closely at what exactly was written in the posts and comments.

A close reading of the commentary section of one entry, "Das nennt man wohl Erziehung," translated by the Knut fan "nene" as "I Guess That's What You Call Education," provides an example of how the meaning of "Knut" was negotiated in the blog. The entry was posted on July 11, 2007, four months after the blog was created, and contains 535 comments, the last one entered on July 14. By this time, a community had started to take form and several of the fans knew one another quite well. The German tabloid newspaper *Bild* had recently announced that the "Knut shows"—where Knut and his keeper Thomas Dörflein played together in the enclosure at the zoo—had been stopped, and Knut was for the first time left alone in his enclosure for more than a couple of hours. This is the English version of Knut's post:

> Guys, guys, I never thought that you'll get at loggerheads with each other because of me. I am just a little polar bear. I think I really have to put a few things straight: of course, friends, of course I will continue to be together with daddy! We don't do any shows anymore, but in the mornings and in the evenings we get together to cuddle and play. Nothing has changed about that. And when daddy has a day off, I'll irk Ronny and Marcus a bit more. So, there really is no reason to worry. And yes, it's true: not everything is always perfect. But that's what changes bring with them. I heard that even human children cry sometimes when they want something, and mommy and daddy don't allow it. (Ehm, pssst, that doesn't mean that I cried!) But I guess that's what you call education. Well, and even though I don't always like it—I have to get through this. Friends, I'll make it!
> Your Knut.

The first responses to this post remarked that it was good to have Knut back again. Knut fan "Dörte" wrote, "How nice that the blog works again, and even nicer that you—dear Knut—yet again is doing well!"[5] In another early comment, "Sanna" wrote, "Good to have you back with us on the blog and thanks for telling us about your feelings. You're the expert, not some silly reporters. You look so happy in Fototanten's[6] photos. . . . See how much we love you. We come to see you and we make websites, blogs, and photo sites about you. We also worry about you so much that we sometimes get too emotional."[7] In a separate entry made on the same day, written under the screen name "administrator," Torsten Rupprich explained the reason why the fans were welcoming Knut back: there had been a server breakdown the previous day and the blog had been closed while being restored from a backup. This also meant that Knut's entry from July 10, 2007, including all comments, had been lost.[8]

Another early comment was from "Eva," who posted a press release she had found in which a person named Frank Albrecht expressed his views on the recent events concerning Knut. Albrecht was an animal activist who ignited the first controversy around Knut by stating that the bear should have been killed with a lethal injection. The claim was first printed in the German newspaper *Bild* on March 19, 2007, and then refuted in the same newspaper the next day, with Albrecht stating that he did not advocate the death of any animal; rather, he wanted to bring attention to the negative effects of hand-rearing animals and to the terrible conditions for animals in zoos in general. In the press release posted on the blog by "Eva," Albrecht was quoted as criticizing the zoo for the sudden separation of Knut from his mother figure, saying that it would cause mental and behavioral disturbances.[9]

Five minutes after "Eva's" post, a person using the screen name "Frank Albrecht" wrote a comment addressed to Knut, though whether this was the same Frank Albrecht who had appeared in the media is impossible to know. In the comment, "Frank Albrecht" mimicked the way of speaking that the fans used with Knut, as if talking to a little child. The first lines of his comment read, "Hello Knut. Your boss (zoo director) doesn't like that people become aware that your enclosure doesn't fit the standards of animal protection laws. If not, he wouldn't have organized to have all my critical comments deleted."[10] His comment implied that there really was no server breakdown, but rather that the entire entry and comments from the previous day had been deleted because the zoo director wanted "Frank Albrecht's" comments removed. "Frank Albrecht" continued addressing the problem of the sudden separation, saying that it would lead to behavioral disorders, and also doubting what "Knut" wrote in the main entry: "And you have also lied to your friends when you say 'And when daddy has a day off, I'll irk Ronny and Marcus a bit more.' Who could even control this, when no one is allowed into your backyard?"[11] He then closed by suggesting that they set up a surveillance camera behind the scenes at the zoo so that the fans could monitor whether Knut was in fact not alone there; and as if to add to the tension, he mentioned that in the past, a baby wolf, a sloth bear, and a hippopotamus had all disappeared from the zoo.

There were generally two types of responses to "Frank Albrecht's" comment. Some did not want to listen to what he had to say, as illustrated by a comment from "Claudia K.": "@Frank Albrecht, sorry, but from your post I assume that you first and foremost don't have a good life, as you always have to criticize, constantly. . . . I'm sorry (but just a little bit)."[12] This was followed up a few comments later by "Kathleen," who wrote in English, "@All. Hey, is this blog going to be about chatting with F. A.?????? No, thanks. Let's hear from everyone with messages for Knut."[13] About half an hour later, after some correspondence in German, "Kathleen" (who does not speak German) asked the other bloggers, "Please summarize what F. A. is honking on about, and what bloggers are telling him in return, and then, I wonder how we can tell him to Get Lost???"[14] These responses are indicative of the temperature of the argument that had been transpiring in the deleted entry from the previous day. There was no longer any interest in even discussing the content of "Frank Albrecht's" comments; the fans just wanted him out of the blog altogether.

On the other hand, there were those who supported "Frank Albrecht" or who at least were interested in the information he provided about the lives of zoo animals. A person with the screen name "Neushoorn" wrote, "Frank Albrecht is no spammer and he is not insane. He is an animal rights activist who poses here and elsewhere pertinent questions on zoo politics vis-à-vis hand raising of wild animals."[15] This was quickly contradicted, however, by an American woman, "Clydene," who pinpointed what was perhaps the central problem for many of the fans: "Mr. Albrecht has made his point over and over again. He is driving people away from the blog. He has been asked to provide background for the information he presents, where is the polar bear Sonja that he talks about. . . . I don't want to continue his misrepresentation of the facts. What has the Berlin Zoo had to do with his accusations? His continued discussion of his topics is SPAM. Unless he can produce documents, bills of lading, receipts, photos, etc. He has been told to leave us alone by more than one regular member of the blog."[16]

In this reading of "Frank Albrecht's" intervention, his comments were perceived as irrelevant—as "spam." "Clydene" further pointed out how "Frank Albrecht" had not backed up any of his claims with evidence, and that in her opinion his claims were not connected to the Berlin Zoo, or more specifically to Knut. Her reference to the regular members of the blog in the last concluding sentence emphasized the communal character of the site; the "regular members" had told him to leave them alone, inferring that their opinions alone should be the ones followed.

Most of the fans, during these exchanges, seemed to want to stop the arguments and get things back to normal. Berlin resident "Liesel" did her best to calm people down. She switched from writing in German to English, and pointed out that she witnessed Knut playing with two zookeepers, Thomas Dörflein and Marcus Röbke, that very same day. She continued, "So, whatever is circulated here, regarding early separation from Knut, Knut home alone, and basically whatever was written in the german gutterpress is wrong. Funny that Germanys biggest tabloid was advised not to run a story tomorrow as it planned. I suppose too many pictures that stae [*sic*] the opposite."[17] In other words, the whole controversy was based on untruthful reports in the newspapers, revealing that "Frank Albrecht" in fact did not know what was going on at the zoo, but rather based his arguments on what he had read in the tabloids.

It did, however, appear that the conflict among the fans had been brewing for some time before "Frank Albrecht" appeared on July 10. On July 9, "Marion K." made an alternative blog that became known as the controversy blog, where those who no longer felt comfortable in Knut's blog could express themselves freely. On July 13, the discussion was turning into self-reflection as the fans were struggling to understand what had happened to their community. "Elke" wrote a long comment to explain her reason for retreating to the controversy blog:

> What disturbed me (and I am not the only one) so much was this "mass hysteria" that suddenly broke out when it became known that the shows were stopped. From where—despite several stories of the opposite by many in the blog—this conviction that the relation T. D. [Thomas Dörflein]/Knut was over came, I have no idea. . . . Since I came back, I have not paid attention to those statements, but only written "Knut geht's gut" [Knut is doing well]. You're quarrelling with F. A. [Frank Albrecht] and saying he shouldn't post anymore. Yet in between you're thanking him for his comments. I also agree that it is possible to see the keeping of zoo animals in a critical light, but that was so stereotypical that it gave me the shivers.[18]

This short excerpt from a very long comment demonstrates some of the main functions the blog was thought to have outside the periods of arguments—functions that were for the most part regained after the anger had subsided. In the first part, "Elke" indicated that distrust had developed among the fans; the Berliners' reports from the zoo were no longer taken at face value. The second part was directed at those who kept feeding "Frank Albrecht," through either quarrelling with him or discussing his arguments. What she expressed here was not so much a wish that people would do as she did, and ignore "Frank Albrecht's" comments, as a dissatisfaction with how fractured the community appeared in their reactions to him. This seems to be the core of the problem, and I will return to this point below.

In the last part of her comment, "Elke" wrote, "Mathias wrote a post yesterday at 14.22 that I found good. But even the separation should be handled by professionals and not our hearts. I still see nothing rash, and find it right to slowly start in July, before Dörflein goes on vacation."[19] This foregrounds the deeply emotional character of the blog, as she admits that the fans argue with their hearts rather than from actual knowledge. It is interesting how she still adds her opinion that the zoo management's solution was correct, inferring that in the blog, emotional judgment has value after all. Picking up on the theme of why people chose to go to the controversy blog, "Gilda" asked whether "Frank Albrecht" was driving people away because he was gaining converts or because people were sick of listening to him. "Eva Gregory" answered, "I think people are leaving the blog not due to converts but due to being sick and tired of seeing F. A. on a blog that is Knut's and represents life and love."[20] In other words, the blog was not a place for criticism against zoos; it was a place for "life and love," a notion that supported "Elke's" comment about fans discussing with their hearts.

"Kathleen" offered another explanation: "I've never worked at a Zoo, I don't know much about the range of normal behavior of young animals, or at all. We have always compared Knut to a child, but when the comparison breaks down, we have these fear days."[21] In the blog, the representation of Knut was indeed similar to that of a human child, and for many, this was what made it possible to feel a connection to him. When the fans lost this reference point, the representation of "Knut" became fractured, revealing many differing views that were not apparent while fans were able to follow a master narrative—and that ended up throwing the community into uncertainty. "Leafpure" expressed some of the same views, writing that "Frank Albrecht's" perspective only "adds to our worries and fears about Knut's future and well being and multiplies them," adding that "I think F. A. capitalizes on that. And that's wrong. He should take it some place else. There are plenty of other animal rights forums and sites."[22] These comments connected to the division that became evident in the reactions to "Frank Albrecht." It is as if the opinions of Knut's fans had become more fractured than "Knut" could contain. When the uniform representation of Knut was challenged by reference to incidents in "real life," the friendly conversation among the fans dissolved into politicized discourse, revealing the difficulty of speaking "for the animal."

Through a discussion that amounts to a total of 535 comments over three days, the community slowly returned to what most of the fans defined as the original state. A person with the screen name "Retta" made a heartfelt appeal that summarized the situation on the morning of July 13: "!All: Historically on this blog, we have chosen to celebrate Knuts life, and the joy, delight, and happiness he brings to all of us. And a very nice result occurred . . . we because joined as a society of 'Knutians' and friendships formed among people from all over the world. Now we are wondering why and how our Knutti[23] blog has changed from the soft, sweet, friendly place it was. As is always the case in life, there is rearely [sic] one cause for any problematic situation like this one."[24]

She continued, suggesting that the problems started when they realized that the zoo staff were only human: "There seems to be politics, jealousy, poor decision making, and poor public communications lurking in the background at the zoo, which we have begun to fear could have a negative effect on Knutti's formation and welfare."[25] Then, she suggested, "a possibly sincere, but opinioned and misguided zealot" arrived, spreading further uncertainty among the bloggers. She emphasized that just because the fans' main purpose was to focus on positive issues connected to Knut, this did not mean they did not care about ethical issues concerning animals both in the wild and in zoos. And this is a key to understanding the blog community: "Our objection to thie [sic] 'intrusion' was simply that the object of this blog was in danger of becoming primarily a forum for argument, when its primary objective had developed into being a forum for celebrating the good in the world."[26] In other words, although the discussions that had been going on revealed that there were in fact a lot of different opinions behind their engagement with Knut, there was little room in the blog community for openly expressing these differences. The blog community was created around a simple representation of Knut as a happy animal, but as the heated discussion shows, the community existed on almost contradictory terms: it consisted of a group of people who self-identified as sincerely devoted to issues of animal rights, but who were also dedicated to keeping these issues separate from their conversations about Knut in the blog. Knut worked as both a symbol and a representation of negative human behavior and human-animal relations, but because his fans wanted these things to change while not wanting Knut himself to change, the actual living polar bear could not effectively intervene in these online crises. The solution, according to "Retta," would be "to step back to our original intent. Let us again make Knut and the goodness he engenders the primary focus here."[27] In other words, the answer was a return to the "fiction" where pseudo-fictional stories told by a happy "Knut" took center stage.

When "Retta" wrote about "original intent" she was referring not only to the community, but also to the blog posts presented by "Knut" himself. Written by "Knut" in first person, the posts were focused narrowly on the polar bear cub and his developing maturity. "Knut" never made any value judgments or expressed political concerns; he never commented on issues of global warming or the status of polar bears in the wild; and he never explicitly stated views about zoological gardens. This was a conscious choice made by the administrators; the blog was originally created to resemble a kind of children's story.[28] However, those who commented and kept returning were adults who were there not only because they thought the polar bear cub was cute and found the posts funny. For many, the initial fascination was connected to their engagement in environmental issues and animal welfare. This is reflected in a comment by an American woman, who writes, "I find it fascinating that so many people gather in this Blog to discuss Knut and environmental issues."[29] In other words, the main topics for the blog were perceived as stories directly concerning Knut as well as wider environmental issues. As "Knut's" posts never contained stories about global warming or its consequences, the fact that this issue was brought in by the fans gives us an indication of the prevailing cultural associations attached to the polar bear. Information about the conditions for polar bears in the wild is presented through links or quotes and is then mostly acknowledged by the rest of the community through comments that say thank you, perhaps with a short comment on the content, such as this comment by "Doro II": "That's great. I particularly liked watching the video. Aren't polar bears magnificent creatures! And to think that we're about to destroy it all. Nature's greatest error: we humans. . . . Just my feelings."[30] Causes of and remedies for global warming were seldom discussed as such, but the fans' recurring acknowledgment that environmental issues were engaging reveals that it was an important part of their Knut fandom. This resonates with the quote by "Retta" above, that the objective of the blog was to celebrate "the good in the world." The fans' concern about the environment and animal welfare was always lurking in the background, but there was a tacit agreement that in the blog these issues should be held separate. The argument that sparked when "Frank Albrecht" arrived, then, serves as an example of how the community fractured when the reality of these issues blended with, and consequently destabilized, the positive, preferred narrative of "Knut."

## Cute Knut and the Climate Crisis

A major force behind Knut's initial fame was undoubtedly his cuteness. Tabloid newspapers explained Knut's appeal with what in German is called the *Kindchenschema*, or "child schema." This term was coined by ethologist Konrad Lorenz to explain why humans often respond with care and protection to animals possessing certain physical features similar to those of human infants.[31] The retention of these features is referred to as neoteny, a term describing not only animal babies but all animals, or even objects that exhibit infant-like features. As a cub, Knut had a big head, fluffy white fur, large black eyes, and a flat, round face, appearing as a model for the term "neoteny." Another physical feature that makes polar bears particularly endearing in this sense is the way they walk—swinging their legs outward and then turning them inward, landing slightly pigeon-toed, making the cubs look cute and clumsy and giving grown polar bears a lumbering gait.[32] In humans, being pigeon-toed is most common in infants and small children.[33]

While the "child schema" seems a fitting explanation for parts of the media hype surrounding Knut, it has some shortcomings when trying to understand the blog community. The similarities between the broader media coverage of Knut and the blog are embedded in the language used to describe and address the polar bear himself. By addressing Knut like one would a child, through simple language and words like "cute," "cuddle," and "little one"[34] and giving him nicknames like Knuti and Knutti, the fans' language use provoked people from outside the fan community, who posted comments in the blog criticizing the fans for being seduced by Knut's cuteness. On April 3, 2007, a person with the screen name "André" wrote, "What's going on here is insane—nothing will surprise me after this. Let's see how long this thing with the nursing dad and the beautiful, cute polar bear will work. In a year at most, we'll hear nothing more from them, because then they'll be replaced by the next hype. And you'll all be moving along as well."[35] The criticism from "André" is based not merely on the conversations in the blog; his reference to "hype" shows that he is judging the blog community against the backdrop of the coverage in tabloid and news media. The fans are accused of being charmed by the ephemeral cuteness of the polar bear cub, which leads to an assumption that their involvement is triggered by the hype, and that their interest will fade and die as the polar bear grows and becomes less generically cute.

The criticism may seem justified, particularly in relation to many fans' conclusions that they are there primarily to celebrate "the good in the world." Yet what is also expressed by the fans, but not reflected in the criticism, is that for many of them the blog communication is only one part of a general interest in animals. What is perhaps harder to understand is the long-term bond that the fans developed with Knut as an individual, a bond that persisted after he matured into a fully grown polar bear. This connection is perhaps better understood as a kind of human-pet relationship. In his article "People in Disguise," James Serpell analyzes the human-pet relationship and presents three theories that offer different explanations for pet keeping. One he calls a "cute response," a theory similar to Lorenz's *Kindchenschema*. According to this theory, people have pets because they feel an urge to protect helpless, cute animals. The second theory is that people keep pets because they are unable to have "normal" relationships with other human beings, and the third is that pets are kept because they provide companionship and an enhanced quality of life for the owners, in addition to, rather than as a replacement for, other humans.[36] For some, the bond with Knut might provide a substitute for human relations. Perhaps of greater importance, as evidenced by the tone among the fans in the blog, Knut quite specifically provided them with a sense of companionship and an enhanced quality of life, not only through the bond that was created between fan and animal but also through the bonds that formed internally among the fans.

The assumption that the fans in the blog are merely succumbing to hype relates to what geographer Jamie Lorimer describes as cuddly charisma. According to Lorimer, cuddly charisma is a form of anthropomorphism. He draws on Owain Jones and Kaye Milton, who both stress that there has to be something recognizable in animals to attract human attention, whether the resemblance of a human face or animals that display "a form of reciprocity to human action or concern."[37] One could discuss whether a polar bear cub's face resembles that of a human, but the relation that developed between Knut and his keeper, or foster father as he was called, undoubtedly displayed reciprocity to human action and concern. The source of Knut's charisma is thus not only his cuteness per se, but also the cuteness as it is played out in relation to human beings, and in particular in relation to the "father," zookeeper Thomas Dörflein (fig. 9.2).

Lorimer also evokes the term "flagship species," which refers to "popular, charismatic species that serve as symbols and rallying points to stimulate conservation awareness and action."[38] The way the fans communicate on the blog makes Knut stand out as a symbol in this sense, but the emphasis on cuteness appears to disqualify any claim to seriousness. This freedom from politics is in fact what most of the fans want for this particular sphere, yet not necessarily what they identify with in a larger context. At the beginning of the twenty-first century, the polar bear is arguably a flagship species, which in many cases is what initially drew the fans' attention to Knut. Yet Knut is at the same time an individual, a named polar bear with a very specific development and a crafted personality that they are able to follow closely.

FIG. 9.2 Knut and zookeeper Thomas Dörflein. Courtesy of Christina M. / cute-crazy-Knut blog.

The narrative of Knut as it was constructed, both in tabloid news media and on the blog, followed the same logic when read by an outsider—it was all about Knut's cuteness. Thus Knut could be seen as an example of the power of visual and shallow representations of animals, what art historian Steve Baker has called disnification.[39] To the fans, however, the two different media—the tabloids and the blog—constitute two very different representations of Knut. "Liesel's" lament in the discussion above, that "basically whatever was written in the german gutterpress is wrong," exemplifies the fans' position: the tabloids present a false picture while the fans are closer to the truth. The difference seems to be difficult for outsiders to comprehend, because the fans' way of relating to and addressing Knut in the blog is so similar to the discourse of the tabloids. Using the same words as their tabloid counterparts, random visitors to the blog interpret the fans' commitment as part of the hype and not as an authentic emotional relation, whether to Knut as an individual polar bear or to broader environmental issues. Instead of simply ignoring the fan community, some, like "André," feel the need to comment in the blog, revealing the discrepancy between the way the fans see themselves and how they are interpreted from the outside. The result can be described as boundary work, in which the bloggers try to explicitly mark themselves off from the media, while the intruders argue that these kinds of emotions are not compatible with an engagement in environmental issues.

The boundary work between the fans and the "intruders" is closely connected to the concept of flagship species. It is not only the hype in itself that is criticized in the blog, but also how the hype constitutes a superficial connection to climate change. Geographer Dan Brockington discusses this sentiment in his book *Celebrity and the Environment*, writing that the use of celebrities as symbols might easily obscure what is really at stake.[40] He cites Katja Neves-Graça, who has claimed that the case of Knut shows "how the signs at work in the climate debate 'iconify complex relationships' and then banalize them into mundane consumption."[41] This criticism captures the superficiality that "André" and others criticize, yet it does not fully capture the essence of the blog community. The childlike, playful discourse in the blog was not a result of the banalization of the fans' relations to climate and animal welfare issues; rather, it was an aside. As "Leafpure" bluntly put it, referring to "Frank Albrecht" above, "He should take it some place else." It was not that the discussion in itself was considered irrelevant, just that it should not be part of this particular sphere.

"André" continued his comment: "Keep on traveling to the zoos in this world with your SUV x5, q7, xc90, Touareg and so on. And never start to think independently."[42] Here, "André" compares the fans' blog activities to driving around in polluting cars while pretending to be concerned about the environment, indicating that they say one thing and do another. If they only had started to "think independently," they would have realized that Knut is just an empty symbol. The first one to answer "André" is "xabbu jun," writing, "@André: I never use a car. I think Knut stands for the problems we have with the environment and nature, and that there always is a way and a future for nature and humankind (it all depends on us). You are totally right, there are other things that could be discussed, BUT KNUT IS KNUT and KNUT IS COOL! My OPINION."[43] This quote echoes the simple joy connected to the community in the blog. According to "xabbu jun," Knut stands for the problems of global warming while also representing a better future, yet this is a discussion to be taken elsewhere—an answer that, while describing the fans' need for separate spheres for these issues, somehow also qualifies "André's" comment about hypocrisy. The importance of the symbolism of Knut is emphasized, but the conclusion, in capital letters, naively insists, "KNUT IS KNUT" and "KNUT IS COOL." There is no way for the critic and the criticized to reach an understanding; they are rather reinforcing their differences.

In the next comment, "trombonewolli" asks why it hurts so much to see people be happy, to which "André" answers, "The childish joy doesn't bother me—what bothers me is that you go to demonstrations against the killing of Knut, but you don't give a s**t about the real problems."[44] In other words, the problem is not the Knut worship as such, but the connection the fans make to environmental issues. "André" thus challenges something that is taken for granted in the community—the very connection that constitutes their group identity—and so the answers from the fans seek to protect their territory by marking off the boundaries between "outsiders" and "insiders," rather than explaining what is going on in order to bring the two opposing groups into accord.

The impact of the blog itself was a topic that arose from time to time among the fans as well. In these circumstances, when the intention was to negotiate meaning from within the community, thoughts were expressed in a friendlier tone. On May 3, 2007, "Afroditi" wrote that she found it great to communicate with such an international group of people about what she called "rescue of animals" and proposed that "Knut rallies human consciousness and help us express our 'positive' thoughts. This small animal is something like a catalyst, what do you think about that?"[45] In an extensive answer, "Sanna" expressed her ambivalence toward this. Her point was not that the fans did not care about the environment, but that the blogging in itself might be futile. "Sanna" touched on the problems of preferential attention being given to cute animals, concerns that connect to the theory about cuddly charisma and flagship species. She stated that "it's not just about saving polar bears, it's also about saving people because the effects of climate change would be so dramatic to everyone of us, or at least our descendants."[46] She expressed discouragement about the reluctance of Westerners to change their lifestyle, yet at the same time hope that the attention Knut was getting, from German Chancellor Angela Merkel for instance, might help raise awareness about global warming.

The difference between the comment by "Sanna" and the ones posted by critics from outside is that, in the comments from outsiders, the criticism was directed at the blog community itself. The critical outsiders grouped the bloggers together with people not willing to change their lifestyles to help prevent global warming, resulting in boundary work where the fans were more interested in making a demarcation between the fan community and the outsiders, rather than discussing the issue raised. Knut and environmental issues are intrinsically connected in the fandom because the fans see Knut not merely as a product of media hype but also as an individual animal and a representative of a species that symbolizes global warming. To the fans, Knut was not just a media construction; he was a creature to which they had a deep emotional attachment, an attachment that simultaneously linked to and provided a break from their concerns about environmental issues. Yet as a unifying force, the representation of Knut in the blog was fragile, as the discussions aroused by "Frank Albrecht" demonstrate. Hence, when the fans' intentions and motivations were challenged, the answers were more concerned with upholding the unity of the community rather than engaging in a conversation with critics. Within the community, the mutual understanding that the commenters were there for the same reasons made it easier to touch on these themes without risking disintegration of their common ground.

The use of the well-known Internet forum code of marking comments with "off-topic" or "OT" when writing posts that do not necessarily agree with the objective of the blog also reveals negotiations over meaning within the blog itself. This "netiquette" was even at one point spelled out by "Hesi": "This is Knut's blog, and everyone can express their opinions here. Therefore I think that the content should be directed at Knut. If this is not the case, most bloggers mark their entries with OT = off topic."[47] What was to be considered off-topic, however, was not always clear. The fans often posted links to stories about animals from other zoos, sometimes marking their posts as off-topic, sometimes not. In one post, "Clydene" asked for information about some polar bear cubs at a zoo in the Netherlands, and the answer to her question was marked as off-topic. "Clydene" then expresses thanks for the answer, marking her comment "OFF TOPIC—SORT OF," adding, "Thank you for the information about the Polar Bears at Ouwehands Zoo, which really isn't off topic as we talk about other polar bears around the world, and sometimes other bears."[48]

Links about environmental issues, however, were never marked as off-topic—an apparent paradox, since they do not explicitly concern Knut and his life in the zoo. Yet although not about Knut per se, discussion of environmental issues does not threaten the positive foundation on which the community was created, as discussed previously. Issues concerning zoos themselves, on the other hand, open up possibilities of questioning Knut's welfare in the Berlin Zoo and thus threaten to break down the fans' common ground. The blog, then, was built on a paradox: it was dependent on a uniform representation of the animal, yet was constantly revealing such a position as difficult to achieve and nearly impossible to uphold.

## Conclusion

The boundary work that was continuously occurring in the blog shows how the fans struggled to uphold a self-organized community created around the mythic construction of a constantly happy "cute Knut." These negotiations relate to connections made in both space and time in the narrative of Knut's development. Although the space of Knut-as-individual is "the zoo," the time-space connected to the polar bear in the early twenty-first century is "Arctic space in a time of melting polar ice." As such, the polar bear has become intimately connected to the events of global warming, because this phenomenon directly and visibly affects its habitat. Thus Knut, as a polar bear, referred to two actual places—the zoo, where he was growing up, and the Arctic, where members of his species are in danger of extinction. The representation of Knut as a blogger online manifests as a third place. "The real polar bear" and the anthropomorphized, "blogging polar bear" were entangled, because they were both situated in the zoo, meaning that the peacefulness of the blog depended on the happiness of the real bear in the actual zoo. This unstable, abstract duality made the foundations of the blog fragile, as exemplified by the unrest caused by "Frank Albrecht." The connection between the blog and global warming was more assured, as it concerned the species itself, and not an individual bear. In this case, it is the species that is endangered; Knut, on the other hand, will always be safe as long as the narrative of the polar bear in the zoo is a positive one.

Because of the increasing focus on conservation issues in zoos over the last decades, Knut as polar bear was immediately interpreted in a larger context of global warming and species extinction specific to his time. Yet as an individual he was part of a long tradition of zoo celebrities and zoo pets that goes back to the emergence of the modern zoo in the nineteenth century.[49] However, whereas the famous zoo animals of the nineteenth century were often species considered exotic because they had never been seen in Europe before, Knut represented a species that was construed as special because of the danger that it would disappear. Over the two-hundred-year history of the zoo, the cultural context of zoo animals has changed from displaying something new to displaying something that might soon be lost. In this setting, Knut's blog was constructed by the RBB within the long tradition of creating zoo pets, highlighting the individual zoo animal; through the open format of blog technology, however, the audience was able to take over the power of definition and bring their cultural references into a narrative that had earlier been controlled by zoo administrators and editors of newspapers. Thus Knut's blog is not new in the sense that it constructs a zoo animal as a personality; rather, the technology of blogs represents something new as it invites the audience to partake in the construction. Knut's blog discloses how the audience interprets zoo animals into a larger cultural context and the extent to which they might bring these animals into their own lives.

The construction of Knut in the tabloid media played on sentimentality, not unlike the construction of Knut in the blog. Yet in Knut's blog, the discussions reveal that there is more to the Knut hype than just an automatic parental response to a neotenous animal. The blog that was intended as a fun place for children evolved into a community of adults who developed an attachment to a bear that extended beyond the "child schema." In a sense he could be compared to Barry the Saint Bernard (see Thorsen, chapter 6), in that both animals had an emblematic identity that was filled with content according to shifting opinions. Knut's blog displays how these opinions existed simultaneously within the same culture, much as did the reactions to *The Watchful Grasshopper* (see Rader, chapter 8). Against all the warnings in the media, Knut's fans continued to love him even after his snout grew long and pointy, and his eyes became smaller relative to the rest of his head; the generic cuteness of the polar bear cub that was evoked to explain the early stages of Knut's popularity proved not to be an accurate explanation of the fans' devotion. This group of fans defined their relationship to Knut as something distinct from that of tabloid media, animal activists, and zoo directors. The blog format thus opened up a new arena for conversations about animals, where ordinary people from all over the world were able to meet and exchange opinions. They distinguished themselves from the "media hype" through their special relation to Knut and to one another, developed through communicating on the blog. Cuteness and the climate crisis were equally important in their communication, as were the simple pleasures of receiving firsthand accounts of the development of a polar bear cub in a zoo from the earliest stages of life to maturity. The blog presented the fans with like-minded people that they would never have encountered elsewhere, and it provided them with a space to let down their guard and cultivate the silly, playful sides of their interest in animals.

What the blog discussions show us is that animals are very much present in urban, Western everyday life, not just as random exotic cuties in the tabloid newspapers, but also as long-term conversation partners and objects on the Internet. Although fictional, the stories "written by" Knut also document how an actual polar bear was growing up among humans in a zoo. Stages of physiological development such as teething and gaining weight, learning to walk, swimming, and standing on two legs, as well as coping with separation from his "father" Thomas Dörflein, taught the fans about animal behavior in a specific setting and gave them knowledge that they would not have acquired, had it not been for it being combined with Knut's generic cuteness and the preexisting concerns about environmental issues that first inspired them to visit the blog. Although the blog was established as a fiction, intended for childish amusement and diversion, the adult fans slowly took over the power of definition and turned the blog into a community of awareness, constantly negotiating the meaning of "Knut" through friendly exchanges, quarrels with annoying "intruders," and disagreements among themselves. The uniform representation of "Knut" that was claimed as the core value of the blog was not stable, but rather was the result of a constant struggle over meaning. The moments in the blog in which the positive representation of "Knut" started to break down reveal different opinions on what a polar bear is and should be—not to mention *where* a polar bear should be—and the many views revealed by the resulting discussions show how the animal matters in different ways, both as an identity marker and as a symbol in the fight against global warming.

## NOTES

1. "Good News—Bad News."
2. The blog was shut down due to the passage of a new German law rendering it impossible for state-run broadcasting companies to include entertainment content on their web pages that did not relate to their TV or radio programs. The blog was available online as an archive for the fans until September 1, 2010, when a new law was passed forcing the RBB to remove the blog from the Internet altogether.
3. The comments posted in English are quoted as they stand, with grammatical errors and spelling mistakes. Of course, a lot of these will have been written by people whose mother tongue is not English, so the mistakes are there to keep the authenticity of the communication. I have translated comments written in German, when noted. I have not quantified my material, but to give some approximate numbers, there are at least fifty different screen names in the commentary sections who returned practically every day, in addition to posters who wrote randomly or only once. All screen names are written in quotation marks. There was no registration needed to post on the blog, which means the same person could technically post under several names. There might also have been a number of people who followed the blog without contributing, known as "lurkers," and there is no way of knowing how many there were.

4. "Hallo, ich bin's, Euer Knut."
5. Dörte, July 11, 2007 (4:45 p.m.), in "Das nennt man wohl Erziehung." My translation.
6. Literally "the photo aunts," a description given to the predominantly female fans in Berlin who went to the zoo every day and submitted their photographs to the RBB to be posted on the blog.
7. Sanna, July 11, 2007 (5:05 p.m.), in "Das nennt man wohl Erziehung."
8. "Es läuft wieder."
9. Eva, July 11, 2007 (4:58 p.m.), in "Das nennt man wohl Erziehung."
10. Frank Albrecht, July 11, 2007 (5:03 p.m.), in "Das nennt man wohl Erziehung." My translation.
11. Ibid.
12. Claudia K., July 11, 2007 (5:06 p.m.), in "Das nennt man wohl Erziehung." My translation.
13. Kathleen, July 11, 2007 (5:18 p.m.), in "Das nennt man wohl Erziehung."
14. Kathleen, July 11, 2007 (5:52 p.m.), in "Das nennt man wohl Erziehung."
15. Neushoorn, July 12, 2007 (12:44 a.m.), in "Das nennt man wohl Erziehung."
16. Clydene, July 12, 2007 (1:34 a.m.), in "Das nennt man wohl Erziehung."
17. Liesel, July 11, 2007 (7:20 p.m.), in "Das nennt man wohl Erziehung."
18. Elke, July 13, 2007 (9:28 a.m.), in "Das nennt man wohl Erziehung." My translation.
19. Ibid.
20. Eva Gregory, July 12, 2007 (5:38 a.m.), in "Das nennt man wohl Erziehung."
21. Kathleen, July 11, 2007 (10:13 p.m.), in "Das nennt man wohl Erziehung."
22. Leafpure, July 12, 2007 (6:59 p.m.), in "Das nennt man wohl Erziehung."
23. "Knutti" is an affectionate nickname for Knut often used by the fans in the blog.
24. Retta, July 13, 2007 (11:12 a.m.), in "Das nennt man wohl Erziehung."
25. Ibid.
26. Ibid.
27. Ibid.
28. Stated by both Torsten Rupprich and Dorotea Horedt during interviews, Berlin, January 19 and 25, 2010.
29. Susan Marie, May 6, 2007 (12:53 a.m.), in "Hurra, ich bin fünf! (. . . Monate alt)."
30. Doro II, September 5, 2007 (7:37 p.m.), in "Konfusius."
31. Lorenz, *Studies in Animal and Human Behaviour.*
32. Ellis, *On Thin Ice.*
33. "Pigeon Toe (In-Toeing)."
34. The German words that appear most often correlate to the English ones: *süss* (cute), *knutschen* (cuddle), and *der Kleine* (little one).
35. André, April 3, 2007 (1:17 p.m.), in "Mein Wochenende." My translation.
36. Serpell, "People in Disguise," 124–25.
37. Lorimer, "Nonhuman Charisma," 919.
38. Leader-Williams and Dublin, quoted in ibid., 923.
39. Baker, *Picturing the Beast,* 174.
40. Brockington, *Celebrity and the Environment,* 129–30.
41. Ibid., 130.
42. André, April 3, 2007 (1:17 p.m.), in "Mein Wochenende." My translation.
43. xabbu jun, April 3, 2007 (1:27 p.m.), in "Mein Wochenende." My translation.
44. André, April 3, 2007 (1:42 p.m.), in "Mein Wochenende." My translation.

45. Afroditi, May 3, 2007 (12:40 p.m.), in "Wenn ich ein Junge werd' . . . When I'll Be Youngster . . . Quand je serais garçon."

46. Sanna, May 4, 2007 (12:33 a.m.), in "Milch macht müde Männer munter."

47. Hesi, September 6, 2007 (12:41 a.m.), in "Konfusius." My translation.

48. Clydene, September 6, 2007 (3:32 p.m.), in "Konfusius."

49. Ritvo, *Animal Estate*, 217. Ritvo mentions a system of "starring" established at the London Zoo with the arrival of the hippopotamus Obaysch in 1850.


## BIBLIOGRAPHY

Baker, Steve. *Picturing the Beast: Animals, Identity, and Representation*. Urbana: University of Illinois Press, 2001.

Brockington, Dan. *Celebrity and the Environment: Fame, Wealth, and Power in Conservation*. London: Zed Books, 2009.

"Das nennt man wohl Erziehung." *Knut's Blog*, July 11, 2007. Site discontinued.

Ellis, Richard. *On Thin Ice: The Changing World of the Polar Bear*. New York: Knopf, 2009.

"Es läuft wieder." *Knut's Blog*, July 11, 2007. Site discontinued.

"Good News—Bad News." *Knut's Blog*, January 14, 2009. Site discontinued.

"Hallo, ich bin's, Euer Knut." *Knut's Blog*, March 6, 2007. Site discontinued.

"Hurra, ich bin fünf! (. . . Monate alt)." *Knut's Blog*, May 5, 2007. Site discontinued.

"Konfusius." *Knut's Blog*, September 5, 2007. Site discontinued.

Lorenz, Konrad. *Studies in Animal and Human Behaviour*. London: Methuen, 1970.

Lorimer, Jamie. "Nonhuman Charisma." *Environment and Planning D: Society and Space* 25, no. 5 (2007): 911–32.

"Mein Wochenende." *Knut's Blog*, April 2, 2007. Site discontinued.

"Milch macht müde Männer munter." *Knut's Blog*, May 3, 2007. Site discontinued.

"Pigeon Toe (In-Toeing): Health Topics." University of Iowa Health Care. http://www.uihealthcare.com/topics/bonesjointsmuscles/bone3447.html. Accessed May 31, 2010.

Ritvo, Harriet. *The Animal Estate: The English and Other Creatures in the Victorian Age*. Cambridge: Harvard University Press, 1987.

Serpell, James A. "People in Disguise: Anthropomorphism and the Human-Pet Relationship." In *Thinking with Animals: New Perspectives on Anthropomorphism*, edited by Lorraine Daston and Gregg Mitman, 121–36. New York: Columbia University Press, 2005.

"Wenn ich ein Junge werd' . . . When I'll Be Youngster . . . Quand je serais garçon." *Knut's Blog*, May 2, 2007. Site discontinued.

## ABOUT THE CONTRIBUTORS

BRITA BRENNA is Professor of Museology at the University of Oslo, Norway. Her research concentrates on eighteenth-century natural history, collecting, and description practices, as well as the long history of exhibitions and display technologies. She is the author of recent articles in *Science, Technology, and Human Values* (2012) and *Science in Context* (2011), as well as a co-editor of the interdisciplinary collection *Routes, Roads, and Landscapes* (2011).

ADAM DODD was most recently a Postdoctoral Research Fellow at the University of Oslo (2010–12). His research is concerned with the cultural history of insect-human relations from the early modern period onward. His book *Beetle* is forthcoming from Reaktion.

GURO FLINTERUD is a Research Fellow at the Department of Culture Studies and Oriental Languages at the University of Oslo. She received her Ph.D. in 2013, with the dissertation "A Polyphonic Polar Bear: Animal and Celebrity in Twenty-First Century Popular Culture." She is currently working on a research project on dog breeding in Scandinavia.

HENRY A. MCGHIE is Head of Collections and Curator of Zoology at the Manchester Museum, part of the University of Manchester. He is an ornithologist, with wide interests in field ecology, historical ecology, and promoting an appreciation of the natural world through museum-based research and education. He is currently writing a biography of prominent British ornithologist Henry Dresser (1838–1915).

BRIAN W. OGILVIE is Associate Professor of History at the University of Massachusetts–Amherst and the author of *The Science of Describing* (Chicago, 2006). His current book project, *Nature's Bible*, examines insects in European art, science, and religion from the Renaissance to the Enlightenment.

KAREN A. RADER is Director of the Science, Technology, and Society Program and Associate Professor of History at Virginia Commonwealth University.

She is the author of *Making Mice* (Princeton, 2004), a history of genetically standardized laboratory rodents, and her forthcoming book *Life on Display* (co-authored with Victoria E. M. Cain and published by Chicago) surveys changing strategies of exhibition and education in twentieth-century U.S. natural history and science museums.

NIGEL ROTHFELS is Director of the Office of Undergraduate Research at the University of Wisconsin–Milwaukee. He is the author of a history of naturalistic exhibits in zoological gardens, *Savages and Beasts: The Birth of the Modern Zoo* (Johns Hopkins, 2002), and the editor of the interdisciplinary collection *Representing Animals* (Indiana, 2002). His current work focuses on ideas about elephants since the eighteenth century.

LISE CAMILLA RUUD is a Research Fellow in the Department of Culture Studies and Oriental Languages at the University of Oslo. Her dissertation, "Doing Museum Objects in Late Eighteenth-Century Madrid," examines how objects at the Royal Cabinet of Natural History (established in Spain in 1771) came into being through intersecting scientific, political, economic, and aesthetic considerations.

LIV EMMA THORSEN is a Professor of the Institute of Culture Studies and Oriental Languages at the University of Oslo. She has authored a cultural history of dogs (*Hund! Fornuft og Følelser*, Pax, 2001), and her current book project focuses on the biographies of some iconic animals on display in the Gothenburg Museum of Natural History: a gorilla, a walrus, a Tonkean macaque, and an African elephant.

# INDEX

Italicized page numbers indicate illustrations. Endnotes are indicated with "n" followed by the endnote number.

Academia de Ciencias, 29
"Account of the Birds, An" (Collett and Nansen), 116–19, *117, 118*
Acheta Domestica, M. E. S. (pseudonym and fictional character), 155, 157–58, *159,* 161, 165, *168*
Adelbulner, Michael, 81, 92
*African Game Trails* (Roosevelt), 60
Akeley, Carl
  elephant skin preservation methods, 69–70
  elephant taxidermy displays and conservation, *70,* 70–71, *72*
  with leopard, 66, *67*
  memoirs of (see *In Brightest Africa*)
  photographs of, *65,* 65–66
Alberti, Leone Battista, 83
Albrecht, Frank, 196–200
Albus, Anita, 85
Aldrovandi, Ulisse, 80, 85
Almagro, Luis de, 24
American Museum, 187n6
American Museum of Natural History (AMNH), 1, *2,* 58, *59,* 176, 177–78
Amundsen, Roald, 110–11
*Animal Matters* (exhibition), 4
animal monsters (deformed animals). *See also* pigs, monstrous
  display purpose, 15–16
  historical perception of, 16
  human monstrosities compared to, 21
  motivations and meanings, 16, 18, 19–20, 21
  natural history museum displays of, popularity of, 15
  normal *vs.* abnormal relationship studies, 21–22
animal rights issues, 183–85, 197–98
"Animals as Objects and Animals as Signs" (University of Oslo research project), 3, 4
anthropomorphism
  of animals through social media, 195–96, 200, 201–2, 209, 211
  charisma of animals as, 204
  of dogs in art, 135
  of insects, 158–60, *159, 161,* 166–68, *168,* 171–73
  pet-keeping theories and, 204
  taxidermy and, 41
Applebaum, Ralph, 1
*Archetypa studiaque patris Georgii Hoefnagelii* (Jacob Hoefnagel, et al.), 79–80, 84, 85, 87
Arctic, 111–14, 115
*Arctic Shipwreck* (Friedrich), 111–12
Armstrong, Isobel, 39, 54
art
  aestheticization of nature, 89–92
  dogs in, 135
  insect illustrations, 77, 79–81, 83–89, *86, 88;* anthropomorphic, 158–60, *159,* 166–68, *168,* 171–73
  insects in, 87
  science linked with, 7
Ash, Edward, 137

Baker, Samuel White, 60–62
Baker, Steve, 205
Barazetti, W. F., 137
Barnum, P. T., 187n6
Barry (Saint Bernard dog)
  background and description, 128, 130, 131
  bicentennial exhibitions featuring, 140
  celebrity of, 128–29, 133, 136–37, 145–46
  cemetery sculpture featuring, 138, 140
  in literature, 135, 136, 138–39
  mythological development of, 133–40
  name origins, 131
  skull of, 140, 141, 144
  skull reconstruction drawings, *142*
  taxidermy museum displays, 133, *134,* 140–45, *141, 142*
*Barry: Eine Hommage an die Nase* (exhibition), 140
"Barry, the Saint Bernard" (Rogers), 135
Basco y Vargas, Jose, 17
bees, 171–74, 187n6
beetles, 86–89, *88*
Benjamin, Walter, 53

Bennett, Tony, 41
"Berg (A Dream), The" (Melville), 115
Bergen Museum
    activities of, 45
    display research and methods, 45–53, *51, 53*
    endangered animals and specimens justifica-
        tion, 44
    founding and history, 38
    funding sources, 56n34
    purpose and mission, 45, 52–53
    research for museum practices, 41–42
    specimen acquisition methods, 42–44
Berger, John, 1, 2, 3, 38
Bergh, Henry, 187n6
Berlin Zoo, celebrity animals at, 192–211, *193,*
    *205*
Bernard, Saint, 129, 139
Bewick, Thomas, 112
bird calls, 119
birds. *See also* Ross's Gull
    literary and visual representations of, 112,
        115–19, *117, 118*
    museum displays of, 47–48, 52
    preservation methods, 106–8, *107, 108*
*Birds of Great Britain,* 115
Blankaart, Steven, 80, 83
blogs, 10, 193–211
"Boatswain, His One Friend" (Byron), 135–36
*Book of Famous Dogs, A* (Terhune), 139
Boston Museum of Science, 179–80
Bown, Nicola, 169
Breyne, Johann Philipp, 93
British Expedition, 114
British Museum (Natural History), 6, 46, 48,
    52, 109
Brockington, Dan, 206
Bronx Zoo, 176
Brunchorst, Jørgen, 45–48, 50
Brú y Ramon, Juan Bautista, 23
Budgen, L. M., 155. See also *Episodes of Insect
    Life* (Budgen)
Bullock, William, 40–41
butterflies, 84–85, *86,* 160–61, *162*
Buturlin, Sergei A., *107,* 109–10, 119
Byron, Lord, 135–36

Campomanes, Pedro Rodríguez, Count of,
    33n38
Carlos III, King of Spain, 18, 20, 24
Carlos IV, King of Spain, 29
Carlson, Allen, 97–98n51
Carlson, Charlie, 181–83
Carroll, Noël, 97–98n51

*Catalogue of Birds in the British Museum* (Saun-
    ders), 115–16
caterpillars, 84–85, *86, 162*
*Celebrity and the Environment* (Brockington),
    206
celebrity animals, 101, 180, 192–211. *See also*
    Barry (Saint Bernard dog)
Chicago Museum of Science and Industry,
    178–79
chick hatchery exhibitions, 178–79
child schema, 202–4
Christianity, 22, 91–93, 95, 133, 137–39, 163
Cimitière des chiens d'Asnières-sur-Seine, Le,
    138, 140
Clark, Kenneth, 135
Clavijo y Fajardo, José, 28
*Coleccion de laminas que representan los animales
    y monstruos del Real Gabinete de Historia
    Natural de Madrid* (Brú y Ramon), 23
Collett, Robert, 116–19, *117, 118*
colonialism, and exploitation, 44
Commelin, Caspar, 81
compassion, as museum display goal, 1–2
conservation
    celebrity animals as symbols for, 202, 204,
        206–9
    of elephant taxidermy displays, *70,* 70–71, *72*
    flagship species, 204, 206
    preservation paradox, 7, 44, 58–59, 68–69
Cosslett, Tess, 157
Crist, Eileen, 154
cuddly charisma, 204, 207
Cuningham, Richard, 62, 63
Curran, Andrew, 16
cyclopism, 22, 23

Daston, Lorraine, 4, 16, 120, 128
Davidson, Peter, 112
Dávila, Pedro Franco, 18–19, 20–21, 23, 25–28,
    32n14
*De animalibus insecta* (Aldrovandi), 80
deformed animals. *See* animal monsters; pigs,
    monstrous
Densley, Michael, 102, 121
Derham, William, 92
Deyrolle, Émile, 44
Dicke, Marcel, 79
dioramas, museum, 41, 65–66, 177
disenchantment, 5, 15–16
disnification, 205
*Dog, The* (Youatt), 133
*Dog in British Poetry, The* (Leonard), 135
"Dog of Saint Bernard's, The" (Fry), 135

dogs. *See also* Barry (Saint Bernard dog)
celebrity and heroism, 136–37, 139–40
literary and visual representations of, 133, 135–37
mythology of breeds, 137
in natural history books, 132–33
Saint Bernard breed of, 130–31, 133, 137
type specimens and breed standardization, 144–45
domestic animals, 129, 132–33. *See also* Barry (Saint Bernard dog); dogs; pigs, monstrous
Dörflein, Thomas, 192, 195, 196, 198, 200, 204, *205*, 211
Dorsey, George A., 50
Dresser, Henry, 109–10
Dürer, Albrecht, 87

*Educated Dogs of Today* (Sanborn), 139
education, 1, 18, 45, 48, 50–52, 64
elephants, 59–62, 69–70, *70, 72*
emotion, as museum display goal, 1
endangerment issues, 7, 44, 58–59, 68–69
Enlightenment, 5–6, 15–16, 157
entomology. *See* insects
*Entzauberung*, 15–16
environmental issues
celebrity animals and, 202, 204, 206–9
preservation paradox and animal endangerment, 7, 44, 58–59, 68–69
*Episodes of Insect Life* (Budgen)
anthropomorphic comparisons, 166–68, *168*, 171–73
author biography, 155
cover, 157–58, *159*
dedication, 163
description and overview, 155–57
frontispiece illustrations, 160, *162*
meaning constructs, 155, 174
narrative images of, 157–61, *159, 162*
reception and reviews, 155–56
structure development, 164–65
themes and purpose, 157, 163–65, 169–70
title page, 158, *161*
ethics
endangerment and preservation paradox, 7, 44, 58–59, 68–69
hunting and specimen collections, 64–68
live animal displays, 183–85
zoos and raising wild animals, 197–98
"Excelsior" (Longfellow), 138
exhibitionary complex, 41
"Experience and Poverty" (Benjamin), 53
Exploratorium, 176–77, 180–87, *182*

Fabricius, Otto, 106
*Farthest North* (Nansen), 116, *117*
*Fauna Groenlandica* (Fabricius), 106
*Fauna Helvetica* (Schinz), 132
Feilden, Henry, 114
Fenton, Robert Hay, 110
Field Museum, Chicago
collectors for, 58
elephant displays and conservation, 5, *70*, 70–71, *72*
Somali wild ass photographs, 65, *65*, 66
*Fighting Bulls, The* (Field Museum exhibition), 5, *70*, 70–71, *72*
Fisher, James, 102, 121
flagship species, 204, 206
Fleischmann, Johann Joseph, 84
Floridablanca, Count of, 20, 24
Flower, William Henry, 48, 50, 51–52
Food for Life Hall (Chicago Museum of Science and Industry), 178–79
Forbes, Edward, 163
"Four Elements, The" (Joris Hoefnagel), 79
*Fram* expedition, 114, 116–18, *117, 118*
Franklin, Alfred, 133
Frévilles, Anne Francois Joachim, 138
Friedrich, Caspar David: *Arctic Shipwreck*, 111–12
Frisch, Johann Leonhard, 81
Fry, Caroline, 135

Gabler, Nicolaus, 90
Gardner, Marshall, 125n62
Gessner, Conrad, 80
Ghiberti, Lorenzo, 83
Giesecke, Karl, 106
Girardin, Carlo Alberto, 139
*Glasarchitektur* (Scheerbart), 37, 38, 39
glass cases
framing materials for, 50
history and purpose, 39–41, 54
legendary dogs displayed in, 133, *134*
for public collection displays, 45–47, 48, 49, 50–51, *51, 53*
as standardized display method, 37–39
glass culture, 37, 38, 39, 53
God, 22, 86, 91–93, 95, 133, 163
Goedaert, Johannes, 80–81, 83, 84, 85, 87
Goode, George Brown, 48, 50–52
Gosse, Philip Henry, 153
Gould, John, 109, 115
Goyeneche Palace, Spain, 18
Graille, Patrick, 16
grasshopper exhibitions, 176–77, 181–87, *182*

Great Saint Bernard Pass, 129–30
Greyfriars' Bobby (dog), 147n31
Guya, Francisco de, 24, 30, 31

Hagner, Michael, 16
Haldane, J. B. S., 86
Hamerton, Philip Gilbert, 133–34
Hartsdale Pet Cemetery, 140
Heeb, Niklaus, 140
Heesen, Anke te, 29–30
Heim, Albert, 137
Henning, Michelle, 41
Híjar, Duke of, 20, 30
Histoire des chiens célèlebres (Frévilles), 138
History of British Birds, A (Bewick), 112
History of British Birds, A (Morris), 115
History of the Birds of Europe (Dresser), 109
Hoefnagel, Jacob, 77, 79–80, 84, 85, 87
Hoefnagel, Joris, 77, 79, 90, 92
Horedt, Dorotea, 194–95
Hospice du Grand-Saint-Bernard, 129–30
human-animal relationships. See also live
    animal displays
  child scheme theories, 202–4
  as display purpose, 2–3
  interactions with nature, as social activity,
    122
  pet ownership, 203–4
  social media and, 193–211
  subjective, vs. objective distancing from
    nature, 9, 154
hunting
  heroic-adventure narratives, 59–63, 68
  philosophies on, 7, 44, 58, 64, 68–69
  tragedy of death narratives, 64–68
Huth, Georg Leonhard, 81

Ibis (magazine), 119
Ignis (Jacob Hoefnagel), 79, 84
In Brightest Africa (Akeley)
  description, 58–59
  elephant hunting narratives from, 59–60, 63
  hunting philosophies in, 64, 66–68
  preservation paradox theme in, 59–60, 68–69
  Somalia wild ass hunting narratives from,
    64–65
infant-like features, animals with, 202–3
Insect Entertainment (Insecten-Belustigung).
    See Monthly Insect Entertainment (Monat-
    lich herausgegebene Insecten-Belustigung)
    (Rösel)
Insecto-Theologia (Insect Theology) (Lesser), 77,
    81, 93

insects. See also Monthly Insect Entertainment
  anthropomorphic illustrations of, 166–68,
    168, 171–73
  display challenges, 7
  in illustration, history, 77, 79–81
  live animal displays of, 176–77, 181–87, 182
  as mechanomorphic objects, 154–55
  popularity in art, 87
"Insect Zoo" (Smithsonian National Museum
    of Natural History), 176
Irmscher, Christoph, 132
Izquierdo, Eugenio, 28, 32n14

Jesse, George J., 136
Jones, Owain, 204
junior assistant programs, 180

Kindchenschema, 202–4
Kingsley, Charles, 115
Kirby, William, 163
Kleemann, Christian Friedrich Carl, 78, 79, 90
Knut (polar bear), 192–211, 193, 205
Koren, Johan, 55n19
Kunstschrank, 40

labeling of museum displays, 46–47, 48, 51, 52
Laboratory of Experimental Biology, 178
Landseer, Sir Edwin Henry: The Old Shepherd's
    Chief Mourner, 147n 31
Language of the Nerve Cells, The (exhibition),
    183
Latour, Bruno, 122
law of priority, 105
Lawrence, Elizabeth Atwood, 2–3
Leonard, L. M., 138
Leonard, Robert Maynard, 135
Leonardo da Vinci, 83
Lesser, Friedrich Christian, 77, 81, 93
Lewontin, Richard C., 122
lighting techniques, 47, 48
Lilford, Thomas Littleton Powys, Lord, 115
Linnaeus, Carl, 19, 32n14, 106
Lister, Martin, 81
Litho-Theologia (Lesser), 93
live animal displays
  development and purpose, 176, 177–81
  grasshopper exhibitions, 176–77, 181–87, 182
  public response and ethical debates, 183–86
London Museum, 40
Long, George de, 116
Longfellow, 138
Lorenz, Konrad, 202–3
Lorimer, Jamie, 204

Lucasa, Joseph Domingo, 17, *17*
Ludwig, Heidrun, 92
Lutz, Frank, 177–78

MacGillivray, William, 104–5, 114–15
Malpighi, Marcello, 80, 94
*Man and Beast Here and Thereafter* (Wood), 136
Manchester Museum, 110
marginalization, 1
Marrel, Jacob, 80
Martínez, Joseph Longinos, 33n15
mechanomorphism, 154–55
Meier, Christoph, 140
Meisner, Karl Friedrich August, 129, 131, 132, 136
Melville, Herman, 115
Menthon, Bernard, archdeacon of Aosta, 129, 139
Merian, Maria Sibylla, 77, 79, 80, 83–84, 87, 92
Merrill, Gilbert, 180
*Metamorphosis* (Goedaert), 84
*Metamorphosis of the Insects of Surinam* (Merian), 79, 80, 83–84, 87
Mey, Johannes de, 81
Meyer, Adolf Bernhard, 50
Milton, Kaye, 204
Mitman, Gregg, 4
Moehring, Paul Heinrich, 82
Moffett, Thomas, 80, 83
Montenegro y Palomo, Paula, 25–28
*Monthly Insect Entertainment (Monatlich heraus-gegebene Insecten-Belustigung)* (Rösel)
 aestheticization of nature, 89–92, *91*
 criticism of, 90, 93, 95
 display techniques, 83–84, 94
 frontispiece illustrations, 90, *91*
 history and reception of, 77, 78, 93
 insect collection and working methods for, 81–83, 93–94
 predecessors and influences on, 79–81
 purpose of, 91–93
 study subjects, 84–89, *86, 88,* 93–94
Morris, F. O., 115
Müller, Johann Christian, 93
Murdoch, John, 107, 110, 113–14
museum displays, overview. *See also specific names of museums and exhibitions*
 early, 132
 glass cases for, 37–39, 45–53, *51, 53*
 international standardization of, 38–39, 52
 live animal, 176, 177–87, *187*
 museum practices and research methods, 42
 public *vs.* private, 48–49
 purpose, 1–3, 48
 specimen acquisition methods, 15, 19, 20–21, 24, 25–28, 42–44
 museum nature, 37, 39, 49
Myers, Doug, 176

Nansen, Fridtjof
 bird preservation and specimens, 108
 museum research and travel, 41–42
 Ross's Gull, first encounter with, 101–2, 113, 121
 Ross's Gull breeding discoveries, 114
 Ross's Gull literary and visual representations in publications by, 116–19, *117, 118*
Napoleon, Emperor, 139
National Museum, United States, 48, 50
National Museums, Scotland, 103–4
natural history, study method options for, 153
natural history books, 132
Natural History Museum of Bern, 133, *134,* 140–46, *141, 142*
*Natural History of Our Frogs* (Rösel), 90
naturalists, 132
nature
 aestheticization of, 89–92
 debates on artistic representation of, 83
 diversity and fundamental order of, 5
 museum, 37, 39, 49
 normalization studies, 21–22
 religion and, 22, 91–92, 95, 133, 137–39, 163
*Naturgeschichte der in der Schweiz einheimschen Säugetiere* (Schinz and Römer), 132
Naturhistorische Museum, Hamburg, 41–42
neoteny, 203
Neri, Janice, 87
Neves-Graca, Katja, 206
Newcomb, Raymond, 108
Newton, Alfred, 109
Noble, Gladwyn Kingsley, 178, 181
Nordenskiöld, Baron Adolf Erik, 116, 121
normalization, 5, 15–16, 21–23, 25
Northwest Passage exploration, 112
Nussbaum, Martha C., 121
Nussbaumer, Marc, 130, 140, 141

objectification, 3–4, 121
objectivity, 154
Ojáncano (Spanish folklore figure), 23
*Old Shepherd's Chief Mourner, The* (Landseer), 147n 31
*Omphalos* (Gosse), 153
Oppenheimer, Frank, 180–81, 184–85, 189n39
Osborn, Henry Fairfield, 58–59

Oslo Zoological Museum, 108
owls, 179–80

Pagenstecher, Heinrich, 42
Park, Katharine, 16
Parry, William, 103
Peale, Charles Willson, 40, 41
Pearce, Susan, 40–41
Penny, Thomas, 80
"People in Disguise" (Serpell), 204
pet ownership theories, 203–4
photography, 65, 65–67, 67, 116, 116
Physico-Theology (Derham), 92
pigs, monstrous
  display practices, 30
  donations and motivations, 15, 19, 20–21, 24,
    25–28
  normalization studies on, 21–23, 24–25
  overview, 5–6
  from Philippines, 17, 17
  popularity of, 15, 16, 29, 30
  veterinarian school studies and displays of, 29
polar bears, 192–211, 193, 205
Poliquin, Rachel, 70
political motivations of specimen donations,
  18, 19–20
Potter, Walter, 41
preservation. See also taxidermy
  of bird specimens, 106–8, 107, 108
  historical definition of, 71
  as justification for killing animals, 7, 44,
    58–59, 68–69
  museum displays and strategies for speci-
    men, 47, 48
public vs. private collections, 22, 23, 48–49

rationalization, 5, 16–17
Raupen wunderbare Verwandelung, Der
  (Merian), 80, 92
RBB (Rundfunk Berlin-Brandenburg), 193–94
Real Colegio-Escuela de Veterinaria de Madrid,
  29
Real Gabinete de Historia Natural, El.
  See Royal Cabinet of Natural History
realism, 4
Réaumur, René-Antoine Ferchault de, 77
Redi, Francesco, 80
religion, 22, 86, 91–93, 95, 133, 137–39, 163
research and research collections
  goals for, 56n40
  as museum goal, 28–29, 48
  as private collection, 22, 23, 48–49
  specimen types for, 49

Researches into the History of the British Dog
  (Jesse), 136
Rhodostethia rosea. See Ross's Gull
Richardson, John, 104–5, 114–15
Rifle and the Hound in Ceylon, The (Baker),
  60–62
Rime of the Ancient Mariner, The, 115
Röbke, Marcus, 198
Rogers, Samuel, 135
Rohrdorf, Hans Caspar, 140
Romé de l'Isle, Jean Baptiste Louis, 32n14
Römer, Jakob, 132
Roosevelt, Theodore, 58, 60, 62–63
Rösel von Rosenhof, August Johann, 78–79,
  81, 83, 91–92, 95. See also Monthly Insect
  Entertainment (Rösel)
Rößler, Michael, 90
Ross, James Clark, 8, 103, 104, 112
Ross's Gull (Rhodostethia rosea)
  Arctic expedition encounters, 111–14
  breeding information, 102
  cultural history and rediscovering, 119–21,
    122–23
  description, 101–2
  discovery and scientific naming, 103–6
  literary and visual representations of, 114–19,
    117, 118
  preservation methods, 106–8, 107, 108
  specimen rarity and value, 102, 109–10
Royal Cabinet of Natural History (El Real
  Gabinete de Historia Natural), Spain
  display motivations and purpose, 18, 19–25,
    30, 31
  donation motivations, 18, 19–20, 25–28, 30,
    31
  donation rejections and social status, 28
  founding and history, 18
  monstrous pig drawings and descriptions at,
    17, 17
  publications on animal monsters by, 23
  public vs. private audiences, 22, 23
  specialization reforms, 28–29, 30
Royal Scottish Museum (now National Museums of Scotland), 110
Royal Veterinary School of Madrid, 29
Rudolf II, Emperor, 79
Rundfunk Berlin-Brandenburg (RBB), 193–94
Rupprich, Torsten, 193–94
Ruprecht, Georg, 141–43

Saint Bernard dogs, 130–31, 137, 140–41. See also
  Barry (Saint Bernard dog)
Saint-Hilaire, Étienne Geoffroy, 32n3

Saint-Hilaire, Isidore Geoffroy, 32n3
Salaura, Judas Tadeo, 28, 30
Sanborn, Kate, 139
San Diego Natural History Museum, 176
San Diego Zoo, 176
Saunders, Howard, 114
Schalow, Herman, 110
Scheerbart, Paul, 37, 38, 39
Scheitlin, Peter, 137–38
Schillings, Carl Georg, 58
Schinz, Heinrich Rudolph, 132
*Schistocerca nitens* (grasshoppers), 176–77,
    181–87, *182*
Schüle, André, 192
scientific discovery, process of, 103
scientific naming, 8, 104–5
Serpell, James, 204
Sharpe, Richard Bowdler, 109, 114
Shaw, Evelyn, 181, 182
Smithsonian Institution, 108, 109, 110, 187n6
Smithsonian National Museum of Natural
    History, 176, 179
Snæbjörnsdóttir, Bryndís, 4
social media, 10, 193–211
social motivations of specimen donations,
    19–20, 24, 25–28
Sociedades Economicas de los Amigos del
    Pais, 27
Soler, Gaspar, 15, 18, 19, 30
South Kensington Museum (*now* British
    Museum), 6, 46, 48, 52, 109
specimens
    bird preservation methods, 106–8, *107, 108*
    museum acquisition methods, 15, 19, 20–21,
        24, 25–28, 42–44
    research, types of, 49
    scientific discovery and type, 103, 104,
        119–21, 144
Spence, William, 163
Spooky (owl), 179–80
*Stag Beetle* (Dürer), 87
Startt, Muhaima, 185
Stebbins, Sara, 92
*Stories and Anecdotes of Dogs* (Wood), 136
subjectivity, 154
superstitions, 23
Swammerdam, Jan, 80, 83, 85

taxidermy
    anthropomorphic, 41
    of birds, 106–7, *108*

dog museum displays, 133, *134*, 140–43, *141*,
    *142*
elephant museum displays, *70*, 70–71, *72*
elephant skin preservation methods, 69–70,
    *72*
for preservation and display, 6–7
reconstruction accuracy issues, 143–44
teratology, 16
Terhune, Albert Payson, 139
*Testaceo-Theologia* (Lesser), 93
*Theatre of Insects* (Moffett), 80
Thomas, Keith, 171
Thorburn, Archibald, 115
"Three Little Dinosaurs or a Sociologist's
    Nightmare" (Latour), 122
Trew, Christoph Jacob, 93
trophy shots, 66–67, *67*

Umlauff, J. F. G., 44
University of Christiania, Norway, 108
University of Oslo, 3, 4

*Versuch einer vollständigen Thierseelenkunde*
    (Scheitlin), 137
*Victorian Glassworlds* (Armstrong), 39
Vienna Natural History Museum, 105
Vincent, Levinus, 81
"Vom Einräumen der Erkenntnis" (Heesen),
    39–40

Ward, Henry, 69
Ward, Rowland, 44
Washburn, Bradford, 179
*Watchful Grasshopper, The* (exhibition), 176–77,
    181–87, *182*
*Water Babies, The* (Kingsley), 115
Weber, Max, 32n2
whales, 42–43, *43*
wild asses, Somali, 64–65, *65*, 66
Wilson, Mark, 4
Wise, M. Norton, 1
women, and early natural science exclusion,
    27–28
Wood, John G., 136, 163

Youatt, William, 133, 140

Zimmerman, Andrew, 54
zoos, 38, 176, 197, 200, 209
Zwinger, Jacob, 80